Making National Energy Policy

edited by
HANS H. LANDSBERG

John M. Olin
Distinguished Lectures
in Mineral Economics

Resources for the Future
Washington, D.C.

Printed in the United States of America

Published by Resources for the Future
1616 P Street, NW; Washington, DC 20036-1400

Library of Congress Cataloging-in-Publication Data

Making national energy policy / edited by Hans H. Landsberg.
      p.   cm.
    Includes bibliographical references.
    ISBN 0–915707–70–5 (alk. paper)
    1. Energy policy—United States—Congresses.   I. Landsberg Hans. H.
    HD9502.U52M347   1993
    333.79'0973—dc20   93–25320
                                                                      CIP

⊚ The paper in this book meets the guidelines for permanence and durability of the
Committee on Production Guidelines for Book Longevity of the Council on Library
Resources.

This book is the product of the Energy and Natural Resources Division at Resources for
the Future, Douglas R. Bohi, director. It was edited by Linda Humphrey and designed by
Brigitte Coulton. The cover was designed by Kelly Design.

RESOURCES FOR THE FUTURE (RFF) is an independent nonprofit organization engaged in research and public education on natural resource and environmental issues. Its mission is to create and disseminate knowledge that helps people make better decisions about the conservation and use of their natural resources and the environment. RFF neither lobbies nor takes positions on current policy issues.

Because the work of RFF focuses on how people make use of scarce resources, its primary research discipline is economics. However, its staff also includes social scientists from other fields, ecologists, environmental health scientists, meteorologists, and engineers. Staff members pursue a wide variety of interests, including forest economics, recycling, multiple use of public lands, the costs and benefits of pollution control, endangered species, energy and national security, hazardous waste policy, climate resources, and quantitative risk assessment.

Acting on the conviction that good research and policy analysis must be put into service to be truly useful, RFF communicates its findings to government and industry officials, public interest advocacy groups, nonprofit organizations, academic researchers, and the press. It produces a range of publications and sponsors conferences, seminars, workshops, and briefings. Staff members write articles for journals, magazines, and newspapers, provide expert testimony, and serve on public and private advisory committees. The views they express are in all cases their own, and do not represent positions held by RFF, its officers, or trustees.

Established in 1952, RFF derives its operating budget in approximately equal amounts from three sources: investment income from a reserve fund, government grants, and contributions from corporations, foundations, and individuals. (Corporate support cannot be earmarked for specific research projects.) Some 45 percent of RFF's total funding is unrestricted, which provides crucial support for its foundational research and outreach and educational operations. RFF is a publicly funded organization under Section 501(c)(3) of the Internal Revenue Code, and all contributions to its work are tax deductible.

Contributors

Peter D. Blair

Assistant Director, Office of Technology Assessment, Congress of the United States

Douglas R. Bohi

Senior Fellow and Director, Energy and Natural Resources Division, Resources for the Future

Hans H. Landsberg

Senior Fellow Emeritus and Consultant-in-Residence, Energy and Natural Resources Division, Resources for the Future

W. David Montgomery

Group Director, Energy Research, DRI/McGraw Hill Energy and Environmental Consulting Practice

Vernon L. Smith

Regents' Professor of Economics and Research Director, Economic Science Laboratory for Research and Education, University of Arizona

Margaret A. Walls

Fellow, Energy and Natural Resources Division, Resources for the Future

Contents

Preface

Energy is ubiquitous. It touches just about every facet of human activity. Everyone is a consumer or producer, or both. Market forces determine much of energy's use, as well as the associated costs and benefits. But energy also generates problems that are beyond the capacity of the market to resolve satisfactorily. Environmental pollution is perhaps the most intricate and intractable among them, as we have learned in the last quarter of a century. Supply assurance in the context of national security is another, and so is our concern for equity. What we call the nation's energy policy attempts to give direction to the production, use, transportation, and distribution of energy to help achieve the array of societal goals in the most compatible ways.

Making National Energy Policy offers a contribution to the understanding of the complexity of a few selected policy issues. The volume presents five lectures originally written for the John M. Olin Distinguished Lectureship Series in Mineral Economics at the Colorado School of Mines during the 1991–92 academic year, and subsequently revised in light of comments and recent developments. Under this lectureship program, the Mineral Economics Department each year invites five or six well-known scholars to spend several days in residence with students and faculty. During these visits, participants present a public lecture on a topic they select within the overall theme of that year's series. For the lectures presented here, the general theme was energy policy in the United States. No attempt has been made to transform these lectures into an integrated whole for publication, nor are they arranged in any particular order.

The Colorado School of Mines is grateful to the John M. Olin Foundation for providing the necessary financial support for the Lectureship Series in Mineral Economics, and to Resources for the Future (RFF) for publishing this volume. We would also like to thank Hans H. Landsberg for kindly agreeing to edit the volume, and Robert Patrick for his contribution in organizing the series.

John E. Tilton
Coulter Professor and Head,
Mineral Economics Department
Colorado School of Mines

Introduction

HANS H. LANDSBERG

*M*aking National Energy Policy, in presenting five public lectures on disparate topics, necessarily excludes consideration of many issues important to its theme. Matters related to energy taxation, for example, are not systematically addressed, nor are price controls, foreign trade, resources and reserves, or government subsidies. Furthermore, the presentations do not always provide the extensive documentation that characterizes scholarly essays as distinct from public speeches.

The five lectures collected here do, however, address areas of substantial importance and provide highly informative and stimulating insights into them. The first lecture conveys a feel for the difficulties encountered in fashioning energy policy by the U.S. Congress, a body that reflects the many facets of energy and its many vested interests. The second lecture makes an important contribution to our understanding of the true meaning of security, a concept or goal that has been a constant in energy policy for decades, but especially since the 1950s. The third presentation also addresses one of the durable issues in energy policy—the relationship between energy and the environment. It does so in a manner that contrasts substantially with conventional wisdom. The fourth lecture takes on the problem of domestic regional conflict, an issue that has been as much a bane of policymakers as any other, with each region attaching primary interest to a different facet of energy—not a surprising feature in a country that spans a continent. Finally, the concluding presentation offers a novel way of dealing with the regulation of electric power generation, a persistent issue that has increasingly moved into the foreground of policy discussion and can be truly said to be "in flux."

The Lecturers

Each of the five authors has a long-standing interest and intellectual investment in energy economics, although in different phases.

Peter D. Blair comes to the energy field from the engineering side. With an undergraduate degree in electrical engineering, he earned a doctorate in energy management and policy from the University of Pennsylvania. While at the University of Pennsylvania, he taught courses ranging from energy management to public policy, and directed research on electric power systems, investment decision analysis, and diverse fields of technology. He continues to be on the faculty as adjunct associate professor. In 1984 Blair joined the U.S. Congress's Office of Technology Assessment (OTA), where he has been broadly responsible for the agency's research on energy and materials. As assistant director he is also responsible for OTA's work on international security. Through his familiarity with both technological and policy issues, Blair is especially well qualified to survey U.S. energy policy.

Douglas R. Bohi, who received his doctorate from Washington State University, developed his expertise on oil import policy as early as 1973 in a congressional investigation. An elaboration of his research led to the book *Limiting Oil Imports*, written with Milton Russell and published by Resources for the Future in 1978. Bohi developed the theme in numerous articles and broadened it to address the role of energy security, and the tools to attain it, in the performance of the economy. His achievements in this area led to a Fulbright scholarship in the Netherlands and a visiting professorship in Australia. His extensive writing complemented his teaching at Southern Illinois University during most of the seventies. In 1978 he joined RFF, where he is now a senior fellow and director of the Energy and Natural Resources Division. Bohi is recognized as a leading authority in the intersection of energy policy and economic growth and performance.

W. David Montgomery has long-standing interest and experience in energy and environmental issues, beginning with his work on marketable pollution permits while he was holding a professorship at the

California Institute of Technology. His reach widened soon to address problems of price formation, the economics of utility regulation, the interaction of energy and environmental phenomena and policies, issues in the regulation of natural gas, the cost of oil imports, and various other energy topics. He has practiced his craft in the public sector, in academic institutions, and in the private sector, in such organizations as the Department of Energy, the Department of Defense, the Congressional Budget Office, Charles River Associates, and Resources for the Future. Currently, Montgomery holds the position of group director in charge of energy and environmental matters at DRI/McGraw Hill. His academic achievements include a doctorate from Harvard, a Fulbright scholarship, and a Woodrow Wilson fellowship. His contribution to this volume draws successfully on the diverse areas of knowledge he has acquired in his impressive professional career.

Margaret A. Walls, a fellow in the Energy and Natural Resources Division of Resources for the Future, earned her doctorate in economics at the University of California, Santa Barbara. Following several years as teaching assistant and lecturer in California, she joined RFF in 1987 to work in the area of modeling petroleum supplies and, allied therewith, issues of controlling air pollution caused by the use of petroleum in transportation. She began publishing in 1990 mainly on oil supply matters, and more recently on the economics of alternatives to oil. Walls's contribution to this volume is a logical extension of her knowledge of the problems of the oil industry and the role of government, leading her increasingly to problems and policy issues that address market externalities.

Vernon L. Smith could be legitimately dubbed an "economic inventor." His name is intimately linked to the "experimental economics" approach that attempts to introduce into the theory of economics such phenomena as auctions and other tools to modify the concept and practices of competitive markets. In this context, regulation and deregulation of utilities are points of intersection with the field of energy, as his contribution in this volume demonstrates. A Harvard Ph.D., Smith has held academic positions in a number of universities, including Kansas, Purdue, Stanford, Brown, Massachusetts, Southern California, and Caltech. At present he is Regents' Professor of Economics and research director of the Economic Science Laboratory at the University of Arizona.

The Lectures

From his special vantage point on Capitol Hill, Peter Blair provides a summary but comprehensive view of current national energy policy issues: what forces underlie the policy discussions, some possible rationales for why policy initiatives have been frustrated, and some thoughts on goals or components of an appropriate strategic framework for implementing policy. He reminds us that energy policy must be viewed and pursued in the context of three overarching areas: economic vitality, environmental quality, and national security. Unfortunately, the author notes, congressional battles tend to be fought over much narrower issues. In developing his story he selects three areas for detailed discussion: oil import vulnerability; electricity supply, demand, and the changing structure of the electric power industry; and energy research and development. Given the fact that writing must cut off at a certain point and that the publication process does not usually keep up with current affairs, the author has appended a brief epilogue to call attention to the most recent developments. Because Blair's paper touches on most points that figure in energy policy debates and also provides an historical setting, it is presented as the opening chapter.

Douglas Bohi's paper gets straight to the question of whether free-market pricing and allocation of energy are in the best interest of energy security, or whether government intervention in private markets is required to protect that interest. The issue is made more difficult because there are wide differences of opinion as to the meaning of energy security. In this exposition, energy security refers to the "economic costs caused by a sudden change in the supply, demand, or market price of energy." Basic to this view is Bohi's statement that "when a disruption occurs in international markets, even complete energy self-sufficiency will not prevent domestic energy prices from following world price levels. . . . Attempts to control the market price and to control imports or exports will not alter these basic facts of life. . . . Such attempts merely hide the true costs, shift the burden to others, and in the process raise the true cost of energy compared to a market allocation." The exceptional case for market intervention is that of market failure, the varieties of which the author discusses (regulated industries, monopoly power, inadequate information, and R&D investment). The paper concludes with a searching assessment of different government intervention tools.

David Montgomery offers a view of the relationship between energy and environmental issues that is far from conventional. While it is true that environmental problems are closely associated with the production and use of energy, it is not true, he argues, that environmental policy exerts a major effect on energy markets, or that, as some maintain, energy policy is in truth a function of environmental policy. Montgomery believes that overall amounts of oil, gas, coal, and electricity consumed have been changed only marginally by environmental laws, although where, how, and at what cost production occurs has been changing. This "decoupling" of environmental policy and energy sources, as the author labels it, is due to market forces. His paper examines in detail the implications of decoupling, how it is accomplished, and what it means. To do so, he outlines the association of pollution with different energy sources, the setup of the market for each, and the environmental policies that most affect the main energy sources, paying particular attention to the Clean Air Act and its amendments. Decoupling exists also, he suggests, in the lack of impact of energy policies on environmental problems (for example, energy conservation does not on the whole produce environmental gains). The paper ends with a detailed description of the possible effects of global warming and policies to address it, such as taxes. This may be the one scenario in which decoupling will be difficult to obtain, and the impact on production and use of the different energy sources could be profound.

Margaret Walls's paper analyzes the choice of governmental jurisdiction for energy security and environmental policies. It examines the economic arguments for and against greater decentralization and concludes that the federal government should take primary responsibility for energy security policies, while several other environmental programs should be left to state and local governments. Walls then discusses the problems that can arise in a federalist system even when it appears that the appropriate jurisdiction is in charge of a particular policy. These problems arise because of conflicting goals across jurisdictions and because of the absence of compensating transfers. Two case studies are examined: federal government leasing of lands on the Outer Continental Shelf for oil and gas operations, and federal regulation of pollution from automobiles. While regional conflict is present in many energy situations, concentration on the large issues of energy security and environmental policies allows the author to be specific and thorough.

Vernon Smith's paper is a fascinating exercise in institutional/ technological inventiveness and speculation. It is highly topical, given the development in this country of a serious debate about restructuring the electric power industry. The basic thesis of Smith's lecture is that electric power production can be deregulated—despite its constituting a natural monopoly—provided the necessary institutional changes are made. Of those, Smith finds the crucial one to be the creation of a "computerized dispatch and coordination center, which allows the coordination advantages of central control to be combined with the information advantages of a decentralized auction bidding market. . . . The key issue in institutional design for the deregulation of electric power is in the specification of property rights." This paper outlines in great detail how the various functions of supplying power would be carried out in such a system. Tightly written and accessible above all to specialists in this field, the paper brings an approach to the topic that should trigger the attention of those whose thinking runs along more conventional lines.

U.S. Energy Policy Perspectives for the 1990s

PETER D. BLAIR

I t is a pleasure to be here. I have interacted with faculty and staff at the Colorado School of Mines for most of my years in government service, as well as before, during my years in academia, and I am very pleased to see some old friends, make new acquaintances, and spend some time sharing thoughts about U.S. energy policy in a rapidly chang-ing world. I have taken it as my charge this evening to provide an overview of current national energy policy issues, with particular atten-tion to the perspectives from Capitol Hill—where I spend most of my time. I hope it comes as no surprise that this charge is a somewhat difficult one these days since, for example, more than 150 bills were introduced into the 102d Congress that aspired to be components of "national energy policy." Moreover, I am sure you have heard and I can confirm that there is often much more heat than light shed on energy policy discussions in Washington.

It is also difficult for me to give you any prescriptions about energy policy given my agency's role in the legislative process. Some of

This paper slightly updates the lecture presented in the John M. Olin Distinguished Lectureship Series at the Colorado School of Mines on September 25, 1991. The views expressed are solely those of the author and are not necessarily those of the Office of Technology Assessment.

you may know about the Office of Technology Assessment (OTA), but for those of you who do not, let me give you just a few words of background. OTA is a small nonpartisan support agency of the Congress, our sister agencies being the General Accounting Office, the Congressional Budget Office, and the Congressional Research Service. OTA is governed by a bipartisan board made up of members of the House and the Senate. This board sets our agenda. Our charge is to look at longer-term science and technology issues at the request of the standing committees of the Congress. OTA is nonpartisan, and we seldom make specific recommendations. Rather, we analyze and articulate the implications of alternative options.

I manage OTA's Energy and Materials Program, one of the nine agency programs. As you might expect, much of the Energy and Materials Program's agenda in the last year has been dominated by items related, in one way or another, to national energy policy as discussions of it have evolved on Capitol Hill. Let me briefly recap some recent events: in 1990 the focus was on environmental concerns, as typified by the debate leading up to the amendments to the Clean Air Act passed by Congress in that year. That bill was arguably the most significant piece of "energy" legislation Congress passed in nearly a decade. In the fall of 1990, energy security concerns accompanying the war in the Middle East returned to the top of the energy agenda as they had two decades before. In 1991 and well into 1992, the recession was foremost in people's minds, but images of the war still lingered somewhat and environmental concerns were being more sharply debated internationally, for example, at the United Nations Conference on Environment and Development convened during the summer of 1992.

The result on Capitol Hill of this sequence of events has been a series of frequent and sometimes chaotic discussions between the administration and the Congress and within the Congress itself. This evening, I had thought it would be useful to outline some of the forces underlying these discussions, some possible rationales for why policy initiatives have been frustrated, and some of OTA's thoughts on the goals of a national energy policy, as well as some components of an appropriate strategic framework for implementing such a policy. First, let me put the current debates about national energy policy into a brief historical perspective.

National Energy Policy: A Historical Note

In 1939 President Franklin Roosevelt appointed a National Resources Planning Board to examine the nation's policy options regarding resources. The board recommended government support of research to promote "efficiency, economy, and shifts in demand to low-grade fuels" and that a "national energy resources policy" should be prepared that would be more than a "simple sum" of policies directed at specific fuels."[1]

The nature of energy policy issues took shape during the Roosevelt years, and in 1945 the U.S. Department of the Interior set forth a collection of principles on which to base national energy policy[2] that included—

1. Use the most economic sources of energy to minimize cost.
2. Use plentiful and nondepleting resources whenever possible in place of scarce and depleting resources.
3. Sources of energy with special characteristics should not be used for purposes for which other, less specialized energy sources are available.
4. The best and most efficient technologies should be used without hindrance.
5. Market stability is essential to properly functioning energy markets.
6. The less labor and capital required to energize our economy, the better for the economy; high levels of employment are promoted by efficiency.

Many of these sentiments were repeated and refined by President Truman's National Security Resources Board in 1947, by the 1950–52 President's Materials Policy Commission (known as the Paley Commission after its chairman, William S. Paley), by President Eisenhower's 1955 Cabinet Advisory Committee on Energy Supplies and Resources

---

[1] *Energy Resources and National Policy*, Report of the Energy Resources Committee to the Natural Resources Committee (Washington, D.C., U.S. Government Printing Office, 1939); also summarized in C. Goodwin (ed.), *Energy Policy in Perspective* (Washington, D.C., Brookings Institution, 1981).

[2] Goodwin, *Energy Policy in Perspective*, pp. 13–14.

9

Policy, by the 1961 National Fuels and Energy Study commissioned by the Senate during President Kennedy's term, by President Johnson's 1964 "Resources Policies for a Great Society: Report to the President by the Task Force on Natural Resources," by President Nixon's 1974 Project Independence Blueprint, by President Ford's 1975 Energy Resources Council reflected in his omnibus Energy Independence Act of 1975, by President Carter's 1977 National Energy Plan, by President Reagan's 1987 Energy Security report, and, of course most recently, by President Bush's 1991 National Energy Strategy (NES). In other words, every president since Truman has formally adopted a national energy policy, albeit with varying degrees of emphasis.

Hence, our discussions today are not new by any means, but we should have learned a few things by now. Let me offer a few observations before I talk specifically about some key features of the current debates. First, it is important to recognize that energy policy per se is not an objective in and of itself. This may sound obvious, but many did not really recognize it in the 1970s, especially those of us energy analysts who were busy constructing models of the U.S. energy system with the rest of the economy considered secondarily, sometimes almost as an afterthought. I hope we have learned our lesson, although I have my doubts—I recently attended a Department of Energy (DOE) review of all the major current energy policy models and was astounded to see how little has changed since the 1970s, with the possible exception that all or most of the models are now operating on personal computers. Our analytical capabilities notwithstanding, a comprehensive, strategic national energy policy must derive its direction from the broader and more fundamental national goals of *economic vitality, environmental quality,* and *national security.* Therefore, as we consider the steps necessary to develop a national energy strategy, it only makes sense to develop that strategy in ways that do the best job of supporting these three and other related goals.

Candidate National Energy Goals and Objectives

As noted earlier, throughout the 102d Congress, members considered the Bush National Energy Strategy and a wide range of other energy-related legislative proposals. These deliberations culminated in comprehensive energy legislation that passed in the closing days of the congressional session. In evaluating the various options reflected in these proposals, it is important to weigh them in the context of the three overarching goals noted above. Some energy options support all three goals, particularly those that improve efficiency of production and use. Others can have conflicting results. For example, increased reliance on coal could cut oil import dependence but exacerbate problems of air pollution and global climate change.

OTA has appeared before many committees of the Congress suggesting that a systematic, comprehensive energy policy would be well served by including a set of specific goals against which the contribution of proposed policy actions can be measured. Some argue that goals and objectives smack of socialism, but we seem to be able to articulate national policy goals for education, crime, health care, defense, and the like. In short, if we do not know where we are headed, the question of how to get there does not matter much. This lack of clear goals and objectives, I believe, is one of the potentially fatal flaws in many of the current comprehensive energy proposals.

The National Energy Strategy

President Bush's energy strategy[3] called for "ensuring the availability of adequate energy at reasonable prices, protecting the environment, maintaining a strong economy, and reducing U.S. dependence on unreliable energy suppliers." In spirit, this collection of goals sounds very much like the overarching goals I listed earlier: economic vitality, environmental quality, and national security.

Many of the recent debates about comprehensive energy policy have focused on much narrower issues. For example, the actions proposed in the National Energy Strategy, instead of being gauged against targets aimed at fulfilling these goals, are deeply tempered by President Bush's stated "keystone of the strategy," namely, reliance on market forces. Some in Congress agree that the NES a priori "market test" is appropriate, even though it excludes many options from considera-

---

[3]*National Energy Strategy: Powerful Ideas for America*, 1st ed. (1991/1992) (Washington, D.C., U.S. Government Printing Office, 1991).

tion, such as energy taxes and efficiency standards for automobiles or appliances. Others feel that the result of such a filter is a narrow set of supply options with little attention paid to demand-side measures except for research and development. Consequently, the debates center on issues for which no analytical basis is included. Indeed, the NES until now has been viewed, at least as it was initially presented, by many analysts as a rough outline at best, and at worst as a kind of cheerleader's pamphlet for market mechanisms. The NES's final set of proposed actions could have emerged with a much more balanced portfolio of supply and demand options had all options been evaluated, compared, and presented against each other in terms of their relative contribution to overall policy objectives.

One last note on the NES: many of the details of the NES analytical process appeared many months after the summary report in a series of Technical Annex documents prepared by DOE.[4] We at OTA and many others have studied the final portfolio of options included in the NES and the process that led to it. As we see more of the details, we see that the divisions and confusions have as much to do with our understanding of the "baseline" projections as with the relative effectiveness and implications of alternative policy mechanisms. One result of this and of the "market test" I referred to above is that the major political battles are likely to occur not in the course of a comprehensive package, but rather in the course of debates over individual elements or pairs of elements, for example, over compromise moves such as automotive fuel economy and the Arctic National Wildlife Refuge or reform of the Public Utility Holding Company Act and least-cost planning initiatives.

As Congress considered the Bush strategy and other proposed initiatives that either complemented or replaced the NES en route to comprehensive energy legislation, a strategic framework with well-defined goals could have helped compare various proposed actions in a consistent, systematic way. In the end, under any circumstances, no one should expect that there are energy "fixes" that are easy or quick. There are no magic bullets. For example, we are likely to succeed in

---

[4]The first released was U.S. Department of Energy, *National Energy Strategy*, Technical Annex 2: "Integrated Analysis Supporting the National Energy Strategy: Methodology, Assumptions, and Results," DOE/S-0086P (Washington, D.C., 1991).

easing U.S. oil import dependence only if we establish long-term efficiency and supply goals and then stick to the plan to achieve those goals through periods of both crisis and calm and through periods of varying oil prices. During the past decade, steady supplies, easy efficiency gains, and a retreat in the price of oil seduced us into largely abandoning efforts to push research in energy efficiency and alternative supplies. Major changes in energy systems—and major changes are what must occur if we are to make a difference—require decades and unwavering commitment from citizens, political leaders, and industry. A great deal of time is required in order to effect a major turnover of the capital stock of energy supply and consuming equipment. Short-term strategies—either to spur production or to curb consumption—are often inefficient and traumatic. Certainly, a sensible, comprehensive energy policy must be responsive to sudden turns of event, but it must be fundamentally grounded in long-term strategies.

Finally, so far the major features of the NES are not all that different from several of the many bills under active consideration, such as S. 1220, introduced by Senators Johnston and Wallop, or the series of bills being considered as H.R. 776, introduced by Congressman Sharp. While I do not have time to address all of the NES or its alternatives in this paper, I would like to touch on several of the issues that are most active right now, namely: (1) oil supply, demand, and long-term oil import vulnerability; (2) electricity supply, demand, and the changing structure of the electric power industry; and (3) energy research and development.

Forces Shaping Energy Policy

Let me recap some of the forces currently shaping energy policy choices, and especially how they have changed since the 1970s.

Global economic growth. The desire to promote economic growth dominates much of the policy agenda. As the world economy becomes more and more internationally connected, this growth has increasingly important implications for resource use, environmental quality, and international trade. Of particular concern is the role of the developing countries, which may account for nearly half of the world's annual energy consumption by the year 2010. U.S. policy interests will increasingly be influenced by trends in the less developed countries.

Environmental issues. Until the Gulf War, environmental issues drove energy policy. One could easily argue that the most significant piece of "energy" legislation in the past ten years was the Clean Air Act Amendments of 1990, which have enormous implications for energy use, including clean coal technologies, existing coal-fired power plants, use of natural gas, alternative transportation fuels, and so on. Local as well as global environmental issues, such as acid rain, urban ozone, global warming, nuclear waste, and others, will continue to be a principal force in energy use and energy policy.

Changing Economic Structure. The United States has undergone substantial changes in economic structure over the past decade. These changes include a changing industrial structure, toward a more service-oriented economy and away from more energy-intensive manufacturing; a shifting trade balance that includes increasing imports of energy-intensive manufactured goods such as automobiles and consumer goods; and changing patterns of final demand that include demographic changes, regional migration, and use of new technology, among others.

Trends in Energy Prices. The recent events in the Middle East have prompted some forecasters to revisit energy price projections, especially those for oil, but many forecasters continue to predict modest increases in energy prices.

Continuing Improvements in Energy-efficient Technology. The 1970s and 1980s "primed the pump" of technology innovation in energy efficiency. Despite low and generally stable energy prices for the past decade, the frontier of energy efficiency improvements continues to expand. Considerable future energy efficiency gains in all sectors of the economy are possible with existing technology, but more substantial gains are available with technologies in development as well.

It is hard to measure the net impact of all these forces, but it is useful to look at a few traditional indicators. For example, a commonly cited statistic, especially among energy efficiency advocates, is that over the period from 1973 to 1985 U.S. energy consumption held virtually constant while gross domestic product (GDP) increased by over

45 percent (see figure I). The one exception was electricity, which is discussed later (see figure 2). Some analysts attribute all of this dramatic decrease in energy intensity to improvements in energy efficiency. This is of course misleading, since many other things were going on in the U.S. economy during that period as well, including (I) a changing industrial structure, away from smokestack industries to services; (2) a changing trade balance, in some sense disguising energy use by consuming products embodying energy consumed abroad to produce those products; and (3) a changing consumer demand owing to many of the factors I cited earlier.

Our own analysis at OTA has confirmed some others that over the last decade two-thirds of the change in U.S. energy intensity can be attributed to energy efficiency improvements (regardless of the motivation—mostly price—for adopting these improvements) and one-third of the shift is due to "other factors," such as changing economic structure—more services, changing final demand, and so forth. Other studies, such as DOE's 1989 study of energy conservation, seem to

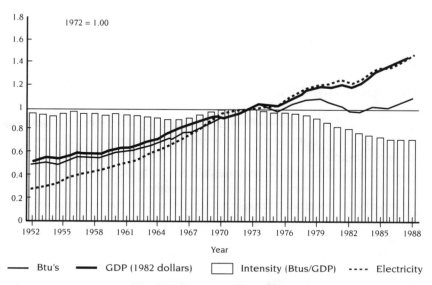

Figure I. Index of U.S. GDP, energy intensity, energy use, and electricity use.
Source: Office of Technology Assessment.

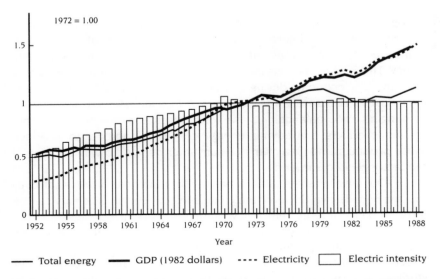

Figure 2. Index of U.S. GDP, energy intensity, energy use, and electricity use.
Source: Office of Technology Assessment.

confirm this and, as Mark Twain said, "A well-informed analyst is one whose views agree with your own."

Oil Import Vulnerability

The 1991 Gulf War focused attention once again on energy security concerns and, in particular, on strategies for reducing U.S. dependence on imported oil.

Today, the world consumes about 65 million barrels of oil per day (MBD). The United States consumes about 17 MBD or about 25 percent of the total world consumption. However, owing to depletion of low-cost resources and lack of new discoveries, the level of domestic oil production is down sharply in the United States. Imports rose from about a third of total U.S. consumption in 1983 to nearly half in 1990. And most forecasts expect up to 65 percent import dependence by 2010 (see figures 3 and 4). Moreover, the percentage of total imports coming from Persian Gulf nations has increased from about 4 percent of total U.S. consumption (10 percent of total U.S. imports) to over 10 percent (26 percent of current total U.S. imports).

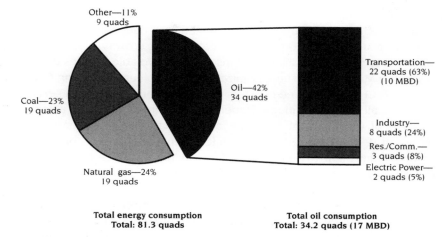

Total energy consumption
Total: 81.3 quads

Total oil consumption
Total: 34.2 quads (17 MBD)

Figure 3. U.S. energy and oil consumption, 1989 (in quadrillions of Btu's [quads]).
Source: Office of Technology Assessment and Energy Information Administration.

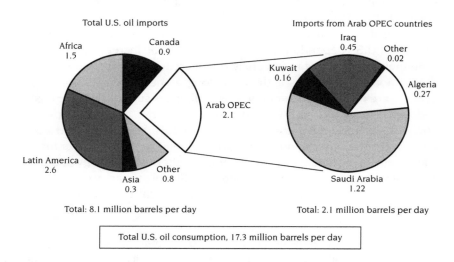

Total: 8.1 million barrels per day

Total: 2.1 million barrels per day

Total U.S. oil consumption, 17.3 million barrels per day

Figure 4. U.S. oil imports, 1989 (in millions of barrels per day).
Source: Office of Technology Assessment.

17

As we have found once again in recent months, diversity and flexibility in oil supply are intimately linked with national security. U.S. energy security is predominately threatened by growing dependence on imported oil, especially on the insecure and volatile sources of supply in the Middle East. The Organization of Petroleum-Exporting Countries (OPEC) controls three-quarters of proved world crude oil reserves, including all recent additions. Dependence on imported oil strains the balance of payments of both the developing and the industrialized world. In 1990 the United States' bill for oil imports was $65 billion, an amount equivalent to well over half of our total balance of payments deficit ($101 billion) for that year. Unless major steps are taken, that dependence will grow substantially.

High levels of oil imports do not by themselves lead to poor international trade balances (Japan is one counterexample). Thus, my economist friends remind me, it is important to distinguish between oil import *dependence* and import *vulnerability.* Dependence is measured simply as the fraction of domestic consumption met with foreign oil. A growing level of imports contributes to import vulnerability, but alone does not translate to vulnerability. Vulnerability arises out of the degree and nature of import dependence, the potential harm to economic and social welfare from a severe disruption in supplies or prices, the duration of such a disruption, and the likelihood of one's occurring.

Because oil use is pervasive and deeply rooted in America's economy, U.S. dependence on oil imports is of increasing concern. Nonetheless, some argue forcefully that increased import dependence should not be viewed as a threat as long as net domestic economic benefits are positive. Current low world oil prices suggest that the costs and benefits should be analyzed carefully. On the one hand, low oil prices have been advantageous for many U.S. businesses and consumers. On the other hand, they have undercut domestic oil ventures, energy efficiency initiatives, and competing alternative energy sources. An honest appraisal of the costs and benefits must take account of all the social, economic, environmental, and political costs of increased import reliance and of the availability of measures to counter the risk it entails.

If it is less expensive (as measured in total indirect and direct costs) to import oil than to offset that need domestically, then it makes sense to import. However, there is strong reason to believe that the reverse is true. In particular, many of the costs of import depe ﹏ce (for example, environmental damage from oil spills and the military costs of protecting oil supplies) are not reflected in the market price of petroleum products. Some of these costs, in particular the military costs, apply disproportionally to the United States relative to the European countries and Japan, all of which have higher energy taxes that move market equilibrium prices for oil use up. Improved energy efficiency, the development of fuels to replace oil, and the reflection of external costs in market prices clearly are in the national interest.

A strategy to limit oil import dependence and vulnerability might involve goals to (1) limit overall oil imports, perhaps to 50 percent of total U.S. oil use, and (2) diversify sources of world oil production and therefore expand U.S. sources of imports to regions of the world outside the Middle East where such imports can be aligned with other U.S. policy interests. Other than noting that the potential for improved production in the Soviet Union, Asia, and South America is great, I shall not address the latter goal here. However, I might mention that OTA completed an assessment of the potential for increased production in the Soviet Union in 1981.[5]

The first goal—limiting imports—could be met with a combination of supply and demand initiatives. The decline in domestic oil production can be slowed with new exploration and recovery technologies applied to existing fields and by opening up new areas for exploration. Developing and producing alternative transportation fuels would shift supplies away from oil. Demand mechanisms include improved efficiency of use in all sectors and shifting of industrial, residential, and commercial oil use into other sources such as natural gas and electricity.

Some options may lead to policy conflicts. For example, development of potential oil resources in the Arctic National Wildlife Refuge, or in sensitive offshore areas, raises serious environmental issues. Similarly, commercialization of technologies for producing alcohol fuels from grain would likely affect food prices and land use patterns. Sustained research and commercial development of new technologies may help resolve these conflicts.

[5]U.S. Congress, Office of Technology Assessment, *Technology and Soviet Energy Availability* (Washington, D.C., 1981).

Figures 5 through 7 illustrate the impacts—in size and timing—of several aggressive strategies in supply, efficiency, and fuel shifting. The options include improving automotive fuel economy, increasing domestic production of oil, switching to alternative fuels in transportation, and a mix of fuel-switching and improved efficiency steps in industry and the residential and commercial sector.

Clearly, as one can see from the figures, vigorous and sustained efforts would be required to hold down oil import dependence over the next several decades—even to a level of 50 percent. The biggest opportunities lie on the demand side. Fortunately, these can provide good jobs and important new economic activity and strength at home. To the extent that we improve efficiency cost-effectively, supplies will last longer, economic competitiveness will improve, environmental problems will be eased, and international tensions lessened.

Improved efficiency, however dramatic, will not be sufficient by itself. The opportunities on the supply side, such as enhanced domestic production in the lower 48 states, offshore, and in Alaska, are more

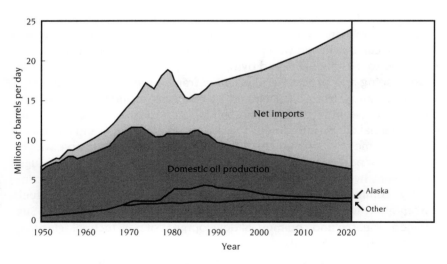

Figure 5. U.S. oil supply and demand futures, baseline projection.
Note: "Other" refers to natural gas liquids and others.
Source: Office of Technology Assessment, adapted from Energy Information Administration, *Annual Energy Outlook* 1990 (Washington, D.C., 1990).

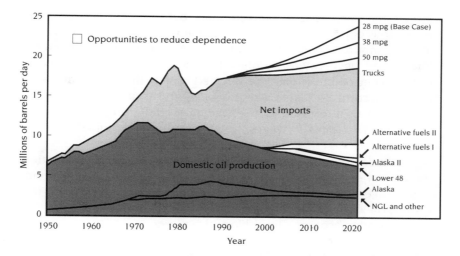

Figure 6. U.S. oil supply and demand futures, increased supply and improved mileage.
Source: Office of Technology Assessment.

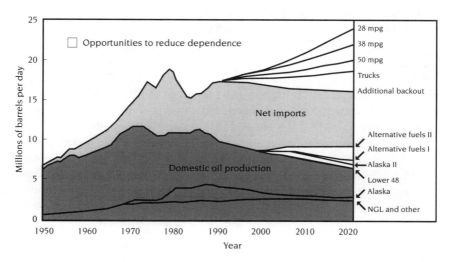

Figure 7. U.S. oil supply and demand futures, improved supply, mileage, and backout.
Source: Office of Technology Assessment.

modest than increased demand efficiency, but still important. And there are various opportunities for using alternative transportation fuels such as methanol, compressed natural gas (CNG), electricity, and others. These fuels have extensive long-term implications, however. The oil replacement potential must be weighed against the energy and environmental costs associated with producing and using these fuels. In the fall of 1990, OTA released its assessment report, *Replacing Gasoline: Alternative Fuels for Light-Duty Vehicles*, which addresses this subject in more depth.

Pacing of all the efforts described above is an essential feature. Patterns of energy supply or demand can change radically as technology changes and as capital stock turns over, but we have learned that short-term changes or policy and technology "quick fixes" can frequently lead to economic hardship and inefficiency.

We focus on three basic scenarios in figures 5–7. The first is a baseline adapted from the Energy Information Administration's (EIA's) *Annual Energy Outlook* 1990 (Base Case).

Scenario I: Baseline. Baseline values for oil production from Alaska and the lower 48 States were extended to 2020 by continuing the 2005 to 2010 trend projected by the EIA. Natural Gas Liquids (NGL) were similarly left at the EIA projected levels to 2010, and then extended to 2020 following the same trend as that projected from 2005 to 2010. The category "Other" (which includes, for example, miscellaneous refinery products) was frozen at the EIA 1995 level of 0.8 MBD; levels in the EIA projection above 0.8 MBD were separated out under the category "Alternative Fuels I" and were added in separately. (See figure 5.)

The EIA projection assumes that U.S. new vehicle fuel economy will reach 38 miles per gallon (mpg) for new cars and 24.4 mpg for new light trucks in 2010—designated the 38-mpg case in figure 6. Our Base Case (figure 5) adjusted the EIA projection to reflect a continuation of the current level for new cars of 28 mpg and for new light trucks of about 21 mpg. Thus the "Base Case" is the 28-mpg case.[6]

---

[6]This resulted in a change in weighted fleet (cars and light trucks) average mileage from EIA projected levels of 23.5 mpg (38-mpg case) in 2010 (on-the-road mileage is assumed to be 80 percent of that measured by the CAFE-type standard) to 20.5 mpg in 2000 (28-mpg

Scenario II: New Supply and Efficiency Improvements. Figure 6 illustrates the impacts of three new sources of supply assumed to begin coming on line in the year 2000 and three levels of transportation efficiency improvements phased in gradually through the year 2020.

The line labeled "Alternative fuels I" assumes that fuels such as methanol and ethanol will be added beginning in 2000 and will reach 500,000 barrels per day by 2020, increasing at a capacity of 100,000 barrels per day every five years.[7] The line labeled "Alternative fuels II" is an accelerated development case.[8] The line labeled "Alaska II" assumes that a large new field is found in Alaska and an accelerated development effort is assumed to result in production levels rising to 500,000 barrels per day by the year 2005.

Figure 6 also depicts three different transportation efficiency improvement strategies, shown as "28 mpg (Base Case)," "38 mpg," "50 mpg," and "Trucks." The 28-mpg and 38-mpg cases were discussed above. The 50-mpg case assumes that by 2020 both new cars and new light trucks have an overall weighted average fuel efficiency of 50 mpg, resulting in an average on-the-road total fleet fuel efficiency of 36 mpg by 2020 (allowing for turnover of the fleet stock and for the 20 percent reduction in actual on-the-road efficiencies from corporate average fuel-economy (CAFE)-type standards). The case for trucks assumes that half of the diesel fuel (heavy trucks, buses, and so

case), which was then kept at this level until 2020. Corresponding annual percentage changes in mileage (on-the-road fuel economy of the total fleet) are 1.18 percent in the EIA projection (38-mpg case) carried through to 2020, and 1 percent in the 28-mpg case until 2000 and then left unchanged from 2000 to 2020. Increasing light-duty vehicle efficiencies are counterbalanced, however, by EIA projections of an increase in Vehicle Miles Traveled (VMT) of 1.82 percent per year from 1988 levels through 2010. This was assumed to continue to increase at this rate through 2020. This increase is driven by such factors as U.S. population growth of 0.8 percent per year (246 million in 1988 to 307 million in 2025) and real GNP growth of 2.4 percent per year overall or about 1.6 percent per year per capita.

[7]These estimates are derived from the EIA Base Case as described above and are comparable to the estimate of the white paper from the Solar Energy Research Institute (SERI), "The Potential of Renewable Energy" (SERI/TP-260-3674, March 1990) for the "Business as Usual" case.

[8]The accelerated development case is listed in the SERI white paper mentioned in footnote 7, which foresees alternative biomass-derived liquid fuels reaching 1.8 million barrels per day more than in the Base Case by 2020 if additional research, development, and demonstration (RD&D) funding is made available.

on) projected to be consumed by 2020 (EIA projection to 2010 extended at the same growth rate to 2020) could be backed out by 2020 through, for example, the use of compressed natural gas and increased efficiency.

Scenario III: Aggressive Oil Backout. Figure 7 includes further oil backout. It assumes (in addition to 50 mpg fuel efficiency and reduced diesel fuel use by trucks) that one-half of the remaining non-transport uses of oil will be backed out in an aggressive oil conservation and fuel conversion program. This results in a savings of 2.7 MBD by 2020, phased in linearly.[9] Compared to current rates of use, this would suggest oil savings in the EIA Base Case of about one-half in the residential and commercial sectors and one-third in the industrial and utility sectors from what it would have been in 2010 if energy use were to grow linearly with economic activity and/or population.

The lesson from these scenarios is that much *can* be done to countervail the ominous projected growth of oil import dependence, but that even with relatively heroic measures we face a future of high dependence on imports. Indeed, the current baseline forecast shows continued declining domestic production and rapidly increasing demand and imports. The principal issue concerning oil use is clearly transportation, which accounts for the largest share of oil use, especially in cars and trucks (60 percent of total use). For the future, substantial alternative fuels, new Alaskan oil production, and various levels of automotive fuel economy improvements will have significant but not anywhere near "energy-independence-style" effects. Even draconian levels of oil backout—replacing half of all other industrial and residential uses of oil—on top of other measures will only hold imports of oil to 50 percent of the baseline forecasted level of total U.S. oil use.

---

[9]The EIA Base Case Scenario already envisages annual residential and commercial consumption of oil decreasing at 2.6 percent and 1.8 percent respectively, between 1990 and 2010, while population grows at 0.6 percent and total real GNP grows at 2.4 percent annually. Oil use in the utility and industrial sectors is assumed to increase at about 0.7 percent per year each, while electricity generation and manufacturing output are assumed to increase 3 percent and 2.8 percent per year, respectively.

Energy Efficiency

Many studies over the past decade have consistently shown that energy efficiency is an essential cornerstone to a comprehensive energy policy framework. As noted earlier, about two-thirds of the falling U.S. energy intensiveness (energy use per dollar of GNP) of the past decade is attributable to improved efficiency in energy conversion and use in every sector of the economy. The other factor prominent in declining U.S. energy intensiveness over this period was the changing structure of the economy, for example, the decline in energy-intensive domestic industries (in many places replaced by energy-intensive imports).[10] The efficiency gains—reductions in energy consumed per unit of service provided (area heated, miles traveled, and so forth)— generally have come about with net cost savings and without sacrifice of comfort or convenience.

Considerable future energy efficiency gains are still possible in all sectors of the economy using existing technology, and even greater cost savings and efficiency gains are possible with technologies under current research and development. For example, in its recent report, *Energy Efficiency in the Federal Government: Government by Good Example*, OTA found that "commercially available, cost-effective measures including high efficiency lighting and carefully operated heating, ventilating, and air-conditioning (HVAC) systems could likely conserve at least 25 percent of the energy used in federal buildings with no sacrifice of comfort or productivity."[11] An efficiency goal of sustained improvement of 20 percent per decade for the next two decades is ambitious but realistic for the United States. A combination of increased research on energy efficiency, increased investment in efficient equipment, and policy leadership can meet or exceed 20 percent improvement per decade at less cost than pursuing the supply-side path. Moreover, pursuing such a goal supports all three policy interests of economic vitality, environmental quality, and energy security.

---

[10]The changing energy structure of the U.S. economy is addressed in detail in OTA's 1990 background paper, *Energy Use in the U.S. Economy.*

[11]This is the first of a series of reports being prepared at OTA in the course of the ongoing assessment, U.S. *Energy Efficiency: Past Trends and Future Opportunities.*

Changing Structure of the Electric Power Industry

I mentioned earlier that the exception to the dramatic drop in energy intensity in the U.S. economy was electricity use. This is due to a number of factors, mostly related to new uses of electricity and the changing market basket of goods and services produced in the United States that I recounted earlier. Aided by stable electricity prices relative to other fuels, U.S. industry, commerce, and residences are becoming increasingly electrified. These changes have contributed to a changing structure of the electric utility industry in the United States that started in the 1970s with the oil shocks. Policy issues center on what, if any, federal role there should be in this changing industry.

The federal government continues to have enormous but diffuse leverage over the nation's generating mix. Federal policy concerns, however, include broader policy interests than those of primary interest to the states. National priorities include security concerns over oil import vulnerability; national environmental, health, and safety regulation; and international competitiveness. It is the sense of urgency about these issues that is likely to dictate the degree to which the federal government plays a role in the next decade. Current programs at the state level, including competitive bidding mechanisms for new supply and demand resources as well as efforts to incorporate demand-side management (DSM) programs and "non-price factors" such as environmental costs into utility integrated resource plans are crucial. They serve both as experiments for increasing competition in the utility industry and as tests of our national ability to internalize environmental and energy security costs into market mechanisms. These state-level programs can be given the chance and encouraged to succeed. If they fail, more aggressive federal regulation will be likely.

At the center of the policy debate over the future of the electric power industry are perceptions of the nations's current and projected electricity generation mix of technologies and fuels through 2010. Historically the mix of capacity and of fuels used in U.S. electricity generation has changed very slowly (see figures 8 and 9). Figure 8 shows the current generating capacity and total primary energy input at electric utilities. Coal is still king, accounting for 43 percent of generating capacity and 54 percent of energy use in utilities, and most forecasts expect continued growth in the use of coal. However, most forecasts also

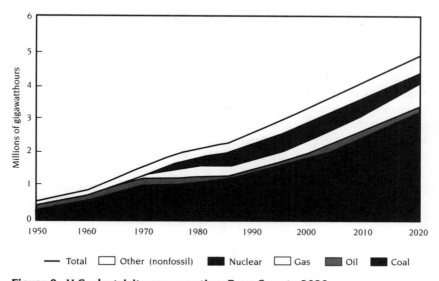

Figure 8. U.S. electricity consumption, Base Case to 2020.
Source: Office of Technology Assessment, adapted from Energy Information Administration, *Annual Energy Outlook* 1990 (Washington, D.C., 1990).

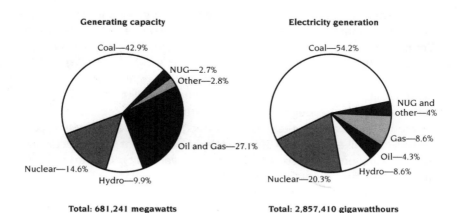

Figure 9. U.S. electric generating capacity and electricity generation by fuel, 1990.
Source: North American Electric Reliability Council, 1990.

27

expect the *share* of coal, nuclear, and hydro power to decline as construction of natural gas and other utility-built sources (solar, geothermal, refuse, waste heat, wood waste, and others) and nonutility generators (NUGs) accelerates.

Much of the generating capacity that will be operating by the
year 2000 is already in place or under construction. The collective
forecasts of the nation's utilities are reported by the North American
Electric Reliability Council (NERC), which projects, for example, coal-
fired generating capacity to increase, but not as quickly as oil and gas.
(NERC expects coal to fall slightly, to 40 percent of total capacity by
2000. See figure 8.) The projections of electricity generated from these
power plants are much more speculative and differ considerably among
various forecasts. For example, the DOE forecast (shown in figure 9)
projects natural gas-fired generation in the year 2000 to be 17 percent
of total generated electricity, while NERC projects only about 9 percent.

In the longer term, through 2010 and beyond, the projections of
both capacity and electricity generation vary dramatically among the
published forecasts. New construction is more important in this time
frame and the dominant fuel of choice for such new generation
depends on expected demand growth, fuel prices, and regulatory policy, all of which are highly uncertain. Hence, the differences among
various long-term forecasts in projected fuel mix result from different
assumptions about total electricity demand (and growth in peak versus
baseload demand), the performance of power plants in terms of efficiency and availability, implementation of the Clean Air Act Amendments of 1990, economic regulatory structure of the electric utility
industry itself (perhaps varying by region), the evolution of technology,
major policy developments in dealing with reducing greenhouse gas
emissions, and, the degree to which natural gas—the current fuel of
choice for new capacity—can be employed aggressively in power generation. The likely fuel mix is particularly dependent on expectations
about future energy resource prices and availability. Concerns about
fuel availability prompt more diversity in resource plans.

Perhaps the largest of these uncertainties is in electricity
demand. For example, to illustrate how badly forecasters have done at
this projection, figure 10 shows the collective forecasts by utilities
through the 1980s, which overestimated demand consistently for many
years to the point that cumulative forecasts represented on the same

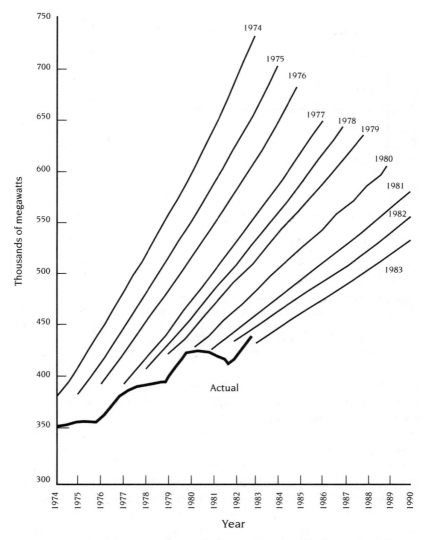

Figure 10. NERC fan—summer peak demand projection for United States, comparison of annual 10-year forecasts.
Source: North American Electric Reliability Council, *Annual Electricity Supply and Demand* reports.

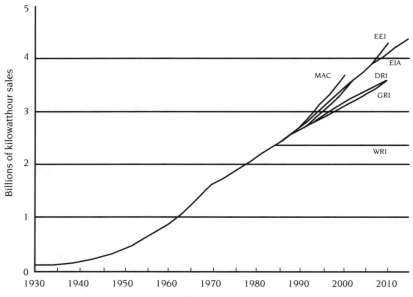

Figure II. U.S. electricity demand forecasts.

Note: MAC = Management Analysis Corp.; Siegel and Sillen; EEI = Edison Electric Institute; EIA = Energy Information Administration; DRI = Data Resources, Inc.; GRI = Gas
Research Institute; WRI = World Resources Institute.

Source: Office of Technology Assessment.

graph became known as the "NERC fan." To be fair, it was not just the
utility forecasts reflected in the NERC compilation that were wrong, it
was those of virtually all other analysts as well.

Figure 11 shows a number of mid-1980 national forecasts by
utilities and others. The primary reasons for the differences in total
demand stem from the different assumptions of the forecasting groups
about regional and national economic growth, changing population,
and demographics. As shown in the figure, the current published forecasts range from about 1 percent to about 4 percent average annual
peak demand growth. Simplistically, this translates to a range of around
a 30-gigawatt surplus to a 280-gigawatt shortfall of capacity beyond
currently planned additions and retirements by 2010. Even within individual forecasts, the range of uncertainty is typically very high. For
example, figure 12 shows NERC's projection of total electricity demand
(summer peak demand) for 1999 with an 80 percent probability band of

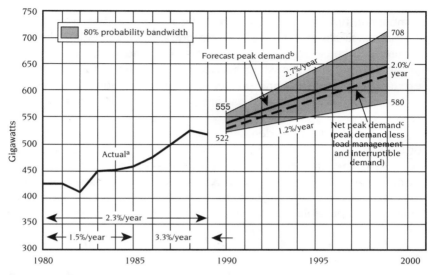

^aUnadjusted for operator-controlled demand reductions, emergency operating procedures, or weather.
^bItem 3A, line 03, April 1, 1990 IE-411 reports.
^cItem 3A, line 06, April 1, 1990 IE-411 reports.

Figure 12. Summer peak demands, United States, 1990–1999, forecasts by NERC.
Source: North American Electric Reliability Council.

128 gigawatts—amounting to a 100-gigawatt shortfall or about a 28-gigawatt surplus at the ends of the uncertainty range compared to currently planned additions and retirements.

I believe that this uncertainty drives much of the decision making in the power industry today. In the old days (1960s for the utility business) of steady demand growth, falling marginal costs (due largely to improving technology), and low interest rates, an excess of new capacity was not all that costly, and demand growth would quickly erase the excess. Now, uncertainty dominates—it is not only greater, but also more important. The tragedy is how long it has taken us to accept any level of uncertainty in forecasting demand and to incorporate it into planning processes. Yet it is now being incorporated in many utilities' planning, as they begin to plan for a range of plausible future scenarios rather than committing to a fixed inflexible plan. When load growth is more sudden than anticipated, as in the New England and Mid-Atlantic regions in the 1980s, shorter lead-time resources such as DSM and

combustion turbines are called upon. Also, some utilities are performing-
ing preconstruction planning and site preparation to reduce the time
required to construct new units, in case demand grows rapidly. The
uncertainty in load growth provides the opportunity to dramatically
expand the role of DSM and smaller-scale, shorter lead-time generating
technologies (such as natural gas-fired combined cycle units) in utility
resource plans.

The uncertainty in demand, supply performance, and availability
of fuels, as well as the decisions that must be made in the face of it,
accentuate the trade-off between minimizing electricity costs and providing adequate generating capacity. Just as the costly excess of capacity that affected many regions in the 1980s was neither planned nor
desired, it is possible that shortages may occur in some regions in the
1990s.

I believe that physical shortages of electricity in the United
States over the next decade are unlikely. Even as structural changes
occur in the industry, the search for a balance between the cost of
additional capacity or of investments to reduce demand and the benefits of higher reliability will remain a driving force in utility planning. The
cost of avoiding such shortages or excesses of capacity could be quite
high, however, unless we can find ways of incorporating efficiency
improvements more effectively into electric utility resource plans. In
the short term, this is the challenge of Integrated Resource Planning.
Without it and more efficient natural gas generating units such as combined cycles, the task of filling unanticipated capacity shortfalls will fall
to much less efficient combustion turbines. As a result, the fuel use and
costs of preserving reliability could be very high. In the longer term, the
role of new technology,[12] strategies for reducing air pollution and
greenhouse gas emissions, and fuel availability will be much more
prominent.

The evolving structure of the industry centers around all the
current trends: independent power, rolling prudence reviews, competitive bidding systems, integrated resource planning, and wheeling

[12]OTA has examined new electric power technologies in *New Electric Power Technologies: Problems and Prospects for the 1990s* (1985), *Nuclear Power in an Age of Uncertainty* (1984), *Starpower: The U.S. and International Quest for Fusion Energy* (1987), among others.

experiments. Many of these ongoing changes are focused on increasing competition in the industry. In its assessment report released in 1989, *Electric Power Wheeling and Dealing: Technological Considerations for Increasing Competition*, OTA concluded that "this assessment has not identified any specific reason why competition cannot be made to work well, but insufficient analysis has been done to determine whether benefits outweigh costs overall. It is clear that there are ways of implementing competition that would work very poorly. There are many pitfalls that must be avoided."

There are opportunities and risks in any approach, including that of retaining the status quo. Of particular concern to us in the current debate over industry structural change is the lack of data and analysis. Any proposed change from the existing system naturally raises uncertainty about how well the new system will work. We know that today's power system works, although some believe it to be somewhat inefficient or inequitable. We also know that the system is currently evolving and accommodating increased competition: nonutility generation and competition among suppliers is increasing substantially in many regions of the country; transmission access is also increasing, although to a lesser degree. However, we will not know the actual impact of these changes on the reliability and economy of the power system for years to come. For the most part, the advantages and disadvantages are still speculative.

Global Warming

In 1991 OTA released its assessment report, *Changing by Degrees: Steps to Reduce Greenhouse Gas Emissions*, which outlines the technical steps that would be necessary to reduce U.S. carbon dioxide emissions. In that analysis we examined several alternative scenarios. In addition to a baseline scenario, OTA constructed a so-called Moderate Scenario that involves measures typically requiring some capital investment but later saving money through fuel savings, which in most cases more than compensated for initial costs. While none of the measures included in this scenario are difficult to achieve technically, inducing consumers to use them may not be easy. A so-called Tough Scenario lowers energy demands further than the Moderate Case, but includes measures that cost more for the same level of convenience or comfort and measures that would be technically difficult to achieve. All of the measures in the Tough Case are technically feasible, but while most are not based on best available prototypes or practices, OTA made judgments about what would be feasible for widespread use. Implementing the technically feasible Tough measures would be challenging—politically, logistically, and economically. The net compliance cost of the Tough Scenario is inherently uncertain but would range from being slightly profitable to costing perhaps $150 billion per year, depending upon a variety of factors such as future energy prices.

Nuclear Power in the United States

The three major impediments to a new generation of nuclear power plants in the United States are commonly cited as (1) licensing reform, (2) advanced reactor designs, and (3) nuclear waste disposal. In 1984 OTA delivered its report *Nuclear Power in an Age of Uncertainty,* which addressed these issues and concluded, "Without significant changes in the technology, management, and level of public acceptance, nuclear power in the United States is unlikely to be expanded in this century beyond the reactors already under construction." We think that this conclusion still stands.

Resolving the issues of licensing reform and nuclear waste are necessary but far from sufficient. The need for advanced reactor designs stems from two principal factors:

1. Scale of Existing Technology. The uncertainty in electricity demand growth over the next decade makes utilities wary of guessing wrong and overcommitting to large, central-station power plants, nuclear or otherwise. Utilities are seeking technologies that are smaller in scale, with modular design features that allow utilities to add generating capability in small increments with short lead times, less concentration of financial assets, and less uncertainty about future regulatory changes.

2. Public Perception of the Safety and Performance of Existing Plant Designs. The actual safety and performance of existing nuclear power plants will probably be debated forever, perhaps with no resolution. The inability to resolve that debate, however, renders the argument irrelevant to the future of existing technology, at least in the United States. Rather, public policy decisions will hinge on the public perception of safety and performance, as it has for the last decade. At this point, that public perception is likely to require new plant designs that incorporate, for example, passive safety and other design features that are not part of existing commercial reactor designs. Public perceptions about nuclear power, however, will be influenced by other issues as well, such as global climate change, continued concerns of air pollution from fossil-fired power plants, and the commercial development of electric generating technologies using renewable energy sources.

The order of resolution of these issues may be very important. For example, a prolonged debate over licensing reform is virtually certain with existing technology. If the nuclear waste issue is resolved and if new reactor designs that respond to public concerns are available and demonstrated, licensing reform debates would likely be more productive.

Finally, many existing nuclear power plants in the United States operate efficiently and have excellent operating records. We do not yet know, however, whether their operating lifetimes can be extended. The issue is very important and worth considerable attention.

Technology Research, Development, and Demonstration

The current U.S. strategy for developing new energy technology assumes that there is adequate time to let private development efforts fill energy supply gaps as they appear. This strategy, however, may not be sufficiently sensitive to the concerns that in the past stimulated special interest in these technologies. In particular, private industry cannot be expected to explicitly include environmental concerns or nonmarket considerations of foreign policy and national security prominently in corporate investment decisions. The federal government plays the principal role in encouraging and sponsoring technology development for such reasons. Of particular concern is assuring availability of liquid fuels as substitutes for oil and improving efficiency in the use of oil, on which virtually our entire transportation system relies. Another concern is finding more environmentally acceptable ways to generate electricity. The current period of low and stable world oil prices (relative to the 1970s) has provided a window of opportunity for developing supply substitutes and more efficient end-use technologies, to ensure commercial availability of these technologies in the 1990s.

In the early 1980s the policy tools thought most appropriate for accelerating commercial availability of alternative technologies were direct energy policy initiatives, such as technology-specific tax incentives, loan guaranty programs, and others. The most effective policy initiatives may no longer necessarily focus explicitly on energy attributes alone, since, as noted earlier, the consideration of such attributes relative to others (such as ease of use or installation, for example) in making investment decisions considering alternative technology and energy efficiency has changed. Indeed, OTA's 1983 study on industrial energy use found that such incentives rarely affected investment decisions.

Perhaps viewing energy alternatives and improved energy efficiency as a collateral benefit to more general economic efficiency has really been the case all along, and policymakers were not sufficiently sensitive to it. Nonetheless, such a perspective ultimately affects the focus of research and development as well, since the kinds of technologies most likely to thrive in such an environment are in many cases different from those most attractive in an environment of urgency about energy. Indeed, decision makers can afford to be more discriminating and demand higher standards of new equipment, thus requiring even more emphasis on commercial demonstration of new technologies.

The fundamental structural changes that have occurred in the U.S. economy over the past decade also have had a profound impact on the relative effectiveness of energy policy tools. For example, the technology-specific initiatives of the 1970s, such as the geothermal loan guarantees or the conservation tax credits, seem strangely anachronistic when compared with more general economic incentives promoting economic competitiveness in world markets that collaterally promote energy efficiency or alternative technology.

Among the most important conditions necessary to sustain energy technology development, especially for conservation and renewables, in the United States is a continued federal presence in research, development, and demonstration (RD&D). While many of these technologies are no longer in the basic research phase, development hurdles are still formidable and the importance of RD&D remains high. OTA's 1985 report, *New Electric Power Technologies: Problems and Prospects for the 1990s*, identified a number of alternative policy options aimed at accelerating the commercial availability of a range of new electric power technologies. These options apply in many respects to other new energy technologies as well, since they focus on reducing cost, improving performance, and resolving uncertainty in both cost and performance. A key to sustained progress in research and development is providing a stable environment for support, so that long-term research ideas can be encouraged, or at least not penalized. Recently, the desire to reduce carbon dioxide emissions has affected the traditional priorities for research and development among energy supply and demand technologies as well.

Renewable Energy

The United States' use of renewable energy as a fraction of total energy use is projected to rise from 8 percent currently to 10 percent by 2010.[13] For decades most people have assumed that fossil fuels can supply human energy needs for several more centuries. The spectre of global climate change as a result of carbon dioxide emissions may invalidate that assumption. The fossil fuel era may have to end, not in centuries, but in a century or less. Nuclear power (fission and fusion) and renewables are the only energy sources that do not emit carbon dioxide. At present, they supply only about 15 percent of our energy needs, but they may soon have to become globally dominant much more quickly than their current pace of development suggests.

Compared with nuclear power, renewable energy technologies have attracted only modest investments in research and development (R&D) from both public and private sources so far. Consequently, major innovations are not apt to come about soon for many of these technolo-

---

[13]U.S. Department of Energy, Energy Information Administration, 1990 *Annual Energy Outlook: Long Term Projections*, DOE/EIA/0383(90) (Washington, D.C., 1990).

gies, compared with the likely incremental changes ahead for nuclear and fossil fuel technologies. On the other hand, while comparative costs still favor fossil fuel technologies, the costs are converging. The cost of renewables is now within a factor or two of that of fossil fuels, down dramatically within the past decade. Energy policy and environmental policy suggest that renewables as well as nuclear energy should be important options, but both will require substantial R&D before they can play significant roles.

Why Have Policy Initiatives Been Frustrated?

Earlier I recounted the view that energy policy should be seen as a derivative or perhaps a supporting policy of broader policy concerns—economic vitality, environmental quality, and national security. Such a view, of course, makes energy policy considerations more complicated and more vulnerable to the political pressures that can frustrate policy initiatives. During a seminar that we convened in Washington last summer,[14] a colleague of mine on Capitol Hill, Jack Riggs, who is the Staff Director of the House Energy and Power Subcommittee of the Energy and Commerce Committee that has jurisdiction in many of these matters, defined seven political system failures that have frustrated energy policy initiatives in recent years. I would like briefly to recap these failures and make one addition.

1. Short Time Horizon. Both our political and economic systems are focused on short time horizons, such as the next congressional election or next quarter's earnings report. Such a focus detracts from considerations about long-term energy policy.

2. U.S. Political System Inertia. The U.S. political system favors the status quo in that it operates under a very conservative constitution with many built-in checks and balances. Such a system requires consensus building that can be difficult in a country with many competing values and interests.

---

[14]U.S. Congress, Office of Technology Assessment, "Environmentally Sustainable Energy Policy: An Exchange of Views on Alternatives," Seminar on American and Swedish perspectives on energy policy issues, May 21, 1991.

3. Disagreement Over the Role of Government. The major political parties as well as other divisions disagree concerning the degree to which government should be activist in energy policy. The shift from activist policies in the 1970s, to laissez-faire policies in the 1980s, to perhaps more moderate policies in the 1990s, has been a hectic ride.

4. Public is Not Knowledgeable About Energy or Environment. Several months ago we were visited by the Gallup Organization with some depressing findings of a poll completed last spring, including the result that over 45 percent of the American public was unaware that the United States must import any oil at all.

5. Energy Choices Heavily Influenced by Price. Even though many of the costs of energy use are not reflected in market prices of energy, price dominates energy use decisions, and uncertainty about price is a particularly important concern.

6. Weak Political Parties. In energy and environmental issues, as with many others, the major political parties are splintered by many coalitions influenced by regional or other common interests that transcend the traditional political party definitions.

7. Competing Values. There are many honest differences among policy interests over the role of energy in our society. These differences are not well defined but frequently surface in debates over energy, environment, and resource issues.

I would like to add one more point, a failure that has to do with government organization. Jurisdiction in the Congress over energy and environmental issues is spread out among many committees and subcommittees. Similarly, energy is a key concern of many executive departments. DOE itself lives a schizophrenic existence between defense and civilian responsibilities, the former of which substantially dominates, at least with regard to the budget. If we add the federal-state jurisdictional uncertainty in many energy areas, the focus of energy policy formulation is diffused across all levels of government.

Concluding Postscript (March 1, 1993)

Since this paper was delivered, comprehensive energy policy legislation emerged from nearly three years of formulation and debate and was enacted into law in the closing days of the 102d Congress as P.L. 102–342. This legislation addressed the widest range of energy policy issues since the early 1970s, including promotion of energy efficiency, nuclear power plant license reform, tax incentives for domestic oil and gas production, sweeping changes in electric power industry regulation, and aggressive promotion of alternative fuels. Nonetheless, this legislation is far from the last chapter of energy policy in the 1990s; numerous issues remain, perhaps most notably fuel economy standards for automobiles and oil production in the Arctic National Wildlife Refuge, both of which are likely to be contentious issues in the next Congress. Moreover, as the Clinton administration took office, the energy and environmental policy perspective of the administrative branch changed considerably. The president's proposed economic stimulus and deficit-reduction package includes a broad-based energy tax and an aggressive government role in the commercialization of new energy technology, almost unthinkable politically just six months ago. Even with this dramatic change in the U.S. energy/environmental political landscape, the trend underscores that today national energy policy is no longer viewed as an end in itself. Rather it must derive its direction from broader and more fundamental national goals of economic vitality, environmental quality, and national security.

Searching for Consensus on Energy Security Policy

DOUGLAS R. BOHI

I t is amazing how fast a war in the Middle East will focus public attention on energy security. Because of Iraq, interest has been revived in an issue that has been dormant in this country for ten years. After several months of timid responses to the question, "what is our energy security policy?," the Bush administration in February 1991 issued the long-awaited National Energy Strategy (NES), giving the government's most recent answer to this question. The program recommends a number of efforts to encourage additional domestic energy production, but no serious proposals to move the country away from imported oil in particular or away from energy in general. Because of these omissions, the NES has provoked considerable controversy.[1]

The debate over a national energy strategy has revealed, once again, the wide differences of opinion, both substantive and ideological, on what constitutes the energy security problem and what the government can do to address the problem. The difference of opinion centers on the question of whether free-market pricing and allocation of energy is in the best public interest, or whether government intervention in

---

This paper was presented in the John M. Olin Distinguished Lectureship Series at the Colorado School of Mines on March 6, 1991.

[1] See, for example, "Cheers and Jeers Greet Energy Plan," *The Energy Daily,* February 21, 1991, p. 2.

private markets is required to protect the public interest. Those with less faith in the efficiency of free markets tend to argue for government intervention to do things that will reduce the economic costs of disruptions in energy markets, and for government policies that over the long term will reduce the nation's dependence on imported oil. Those with an opposing view on the efficiency of markets tend to be skeptical of the benefits that government intervention can achieve, and argue that the costs imposed on the economy by government intervention tend to be hidden and shifted to others, and that the costs frequently outweigh the benefits.[2]

If you were not already skeptical of the possible contribution of the government in addressing the energy security issue, a brief look at past efforts is sobering. Not so long ago, we had President Carter extolling the virtues of the "moral equivalent of war," where the government got in the business of controlling thermostats and promoting sweaters in place of heat. And don't forget President Nixon's Project Independence, launched in 1973 to meet the nation's energy needs without dependence on any foreign sources. The legacy of both programs is a long list of enormous inefficiencies, including natural gas curtailments, gasoline lines, a bankrupt Synthetic Fuels Corporation, a bewildering array of prices for oil and gas that depended on the individual well they were drawn from, and numerous restrictions on the uses of different fuels that made no sense. No one has calculated the losses in productivity forced on the economy since 1973 by these measures, but anecdotal evidence suggests they are very large indeed.

Going back a little farther, remember that President Eisenhower treated the country to oil import quotas in 1959 that limited imports to 12.2 percent of domestic production in order to protect the domestic industry from cheaper imports. That program, which remained on the books until 1973 but actually died years earlier, demonstrated how hard it is to try to separate the domestic market from the international

---

[2]The free-market ideology has occupied the White House for the past decade, and this view is reflected again in the Bush administration's National Energy Strategy. The opposing ideological view, expressed in a *Washington Post* editorial of February 12, 1991, argues that "despite two decades of painful experience, President Bush's White House insists on leaving energy policy to what it calls the free market. . . . Ideology is always the death of sensible policy and particularly in a field as highly polarized as energy."

market. The program had to be continually changed to plug gaps opened by creative businessmen and to allocate the largesse created by an artificial distinction between domestic and world prices.

The price structure protected by the import quota program actually started three decades earlier with the demand prorationing programs implemented by the major oil-producing states. Led by the Texas Railroad Commission, the oil-producing states effectively limited total U.S. oil production to an amount that would just cover demand at a predetermined price. The federal government ratified this approach to controlling the market by passing the Connally Hot Oil Act, which prohibited interstate sales of crude oil produced in violation of state restrictions. When oil imports began to threaten this system in the 1950s, the import quota program was imposed to maintain the status quo.

Therefore, for six decades, until 1980, the U.S. oil market was controlled by one form of government regulation after another. We have had, in other words, one energy security policy after another, usually aimed at protecting the domestic industry from competition by imports. None of these policies contributed in any real way to national security, but all of them left a legacy of waste and inefficiency. In view of this record, what should we expect from the government today? The best advice I can give is old advice: above all, do no harm. In what follows I will try to describe what this entails.

The Role of Government in Energy Security Policy

It is useful to begin by being more specific about what is meant by energy security. In this paper the problem of energy security refers to the economic costs caused by a sudden change in the supply, demand, or market price of energy.[3] For practical purposes, a disruption in the world oil market is the only example of an energy emergency in which the capability of the domestic or international energy system is seriously taxed.

---

[3]Military and diplomatic costs are purposely excluded, although it is sometimes argued that these issues arise only because of the significance of the economic costs. In contrast, the U.S. military response to the Iraqi invasion of Kuwait was, according to President Bush, motivated by several factors in addition to the economic importance of oil.

Before one assigns a specific role for government policy, it is useful to recognize that some costs of an energy emergency are unavoidable no matter what the government does. An increase in energy scarcity, no matter what the cause, means that society must pay a higher price for energy and must learn to use less of it. The higher price may be reflected in the cost of energy imports, when a larger amount of exports is required to pay for the same amount of energy imports, or in the cost of domestic energy production, when more of the nation's resources are required to produce the same amount of domestic energy output. Some price increases may not reflect underlying changes in scarcity, and some adjustments to price shocks may be unnecessary, but fewer resources mean that someone has to do without.

When a disruption occurs in international markets, even complete energy self-sufficiency will not prevent domestic energy prices from following world price levels. Indeed, unless the domestic price rises to the world level, domestic producers would prefer to export their products rather than sell them at lower prices in the domestic market. Attempts to control the market price and to control imports or exports will not alter these basic facts of life, as we have seen before. Such attempts merely hide the true costs, shift the burden to others, and in the process raise the true cost of energy compared to a market allocation.

It is important to recognize, in addition, that the private sector will adjust to the risks of energy disruptions, and that these adjustments will displace the need for government action. Energy emergencies impose potential costs and offer potential profit opportunities that prompt individuals to anticipate and prepare for them. These adjustments involve changes in the way energy is produced, consumed, and stored, and they usually entail costs of their own. Individuals will therefore strike a balance between the costs of preparing for an emergency and the perceived benefits that will be gained by being prepared. Achieving this balance recognizes that buying too much security can be as wasteful as being underprepared for an emergency.

While private preparations are motivated by self-interest, their collective effect is to reduce the demand and increase the supply of energy in an emergency. These actions will reduce the potential cost of an emergency and, in the process, reduce the need for government action. These private adjustments also imply that the risk of a disruption in energy markets does not alone justify government intervention. The rationale for government action must be based on some deficiency in private responses to the risks.

In the process of trying to fill a gap in private responses, there is a danger that government intervention will offset or reverse desired private actions. This can happen because the effect of any government action that is intended to reduce the cost of a disruption will also reduce the private gain attached to being prepared for a disruption. The reduction in both private risks and gains will reduce the incentives for undertaking emergency preparedness. Thus, government actions will displace private actions. This displacement must be taken into account in assessing the desirability and the cost of government action. For example, the creation of the Strategic Petroleum Reserve (SPR), in which the government stores oil to be sold in an emergency, reduces private incentives to store oil for similar contingencies. To the extent that the SPR displaces private storage, the SPR makes less oil available in an emergency than is actually stored in the reserve. It also follows that the cost of the SPR to society is greater than the amount spent on buying and storing oil for the SPR.

A more serious example of how government actions can distort private incentives is provided by the spate of legislation introduced (but never passed) after August 2, 1990, intended to confiscate profits earned by oil companies as a result of the rise in oil prices following Iraq's invasion of Kuwait.[4] Although war profiteering evokes evil visions, those same profits provide incentives to the private sector to take actions that are in the public interest, such as the incentive to store oil that could be used in an emergency, or to draw down available inventories during an emergency to ease the shortage. The confiscation of capital gains acts like a 100 percent capital gains tax on oil, and one can easily imagine the destructive effects on investment incentives if such a tax were to be applied to home sales or stock transactions.

---

[4]The numerous bills introduced in Congress after August 2, 1990, included The National Emergency Anti-Profiteering Act of 1990 (S. 3035 and H.R. 5582), The Iraqi Anti-Aggression and Windfall Confiscation Act of 1990 (H.R. 5580), The Windfall Profits Tax and Deficit Reduction Act of 1990 (H.R. 5758), and An Amendment to the Internal Revenue Code Reinstating the Windfall Profits Tax on Domestic Crude Oil (H.R. 5912).

To properly aim government policy initiatives at avoidable costs and at the same time avoid duplicating or displacing desirable private sector actions, energy security policy should be designed to address the deficiencies in private sector planning and action. When the private sector fails to take actions that are in its interest, there is usually some form of market failure, or what economists call an externality that accounts for the deficiency. Market failures could explain why the private sector might systematically underestimate the prospects for an energy disruption, might fail to recognize all of the costs of a disruption, or might lack the incentive to take actions that will guard against the risks of a disruption. This approach to policy analysis takes the view that the demonstrated existence of a market failure is a necessary (though not sufficient) condition for government intervention. The specification of the target for government policy also serves to focus attention on policy options that are capable of addressing the source of the problem.

Market failures could be responsible for (1) inadequate research and development (R&D) needed to diversify sources of supply and to create substitutes for consumption; (2) the regulation of energy markets and other sectors of the economy that increases the cost of adjusting to energy disruptions; (3) the presence of market power that enables buyers or sellers to control the price of energy; and (4) wage and price rigidities that increase macroeconomic adjustment costs. Each of these topics will be discussed in turn.

Information Externalities

The standard argument that the private sector tends to underinvest in the collection of information may be applied to information required to assess the risks of energy disruptions and to respond to those risks.[5] The argument is that information is costly to gather and assimilate, yet the benefits of information cannot be appropriated exclusively by those who bear the costs. Information is like a public good in the sense that, once provided, it is difficult to exclude individuals from sharing its benefits. As a result, the benefits of additional information tend to be larger

---

[5]For a general discussion of the literature, refer to R. Cornes and T. Sandler, *The Theory of Externalities, Public Goods, and Club Goods* (Cambridge, Cambridge University Press, 1986).

for society than for the individuals who create it. This means that the private sector will invest in less information than is socially desirable. The gap is the result of a market failure, and closing the gap provides a reason for the government to invest in information.

This argument is fine as far as it goes, but it does not say anything about how much of what kind of information the government should support. The distinction is important because the government already generates, collects, and disseminates a great deal of information. One category of government information comprises all of the basic data currently available on energy production, consumption, and pricing of energy, as well as analyses of these data. Another category comprises information that results from government support of R&D.

With regard to the energy security problem, it is probably beyond the capability of the government to help the private sector assess the risks of disruption, except to provide market data that will enable individuals to draw their own conclusions. The reason for doubt is that the government is not a player in the market, and staff members will not win or lose on the basis of the risks and rewards that depend on understanding market behavior. It should not be surprising, as was experienced with previous oil market disruptions, that the government consistently turns to the private sector for understanding about the state of the market and the prospects for future developments.

In contrast, the government can play an important role in supporting R&D that will ease disruption risks. In particular, R&D should be targeted toward diversifying the available sources of energy supply and the range of substitutes for energy in consumption. Enhanced diversity of supply will reduce the significance of any one fuel or any one geographical source of supply and will mitigate the transfer of localized disruptions to the entire energy system. Expanding the range of substitutes in consumption works to reduce the burden on consumers that is caused by any remaining price volatility.

Economic Regulation of Energy Industries

Economic regulation of natural gas and electricity is intended to exploit economies of scale in the production and distribution of these commodities. Competition, in contrast, is wasteful in situations where a single firm can supply the market at lower unit cost than a number of competing firms that divide the market. The inefficiency of competition is the market failure that justifies the creation of legal monopolies. The establishment of legal monopolies, in turn, requires regulation to ensure that the firms operate in the public interest. In return for exclusive franchise rights, the regulator fixes the price of the commodity to cover costs and to allow for a fair rate of return on investment.

Regulation contributes to the energy security problem because firms in these industries do not respond to market incentives in the same way as do firms in competitive industries.[6] Regulation creates built-in rigidities that make it difficult for firms in these industries to respond to the changes in market conditions that are caused by a disruption. Because prices in these markets are fixed by regulatory rules, they cannot be changed except by formal review, and they cannot serve the normal function of allocating scarce supplies to their most productive uses or of encouraging energy producers to increase output.

The lack of adjustment in the natural gas and electricity industries increases the burden of adjustment in the oil market and makes the oil price even more volatile. As a consequence of these rigidities, the economy is unable to adjust quickly and fully to a disruption, making the economic costs of a disruption larger than necessary. These economic costs will show up in the form of lower productivity, higher unemployment, and lost production.

Over the past decade, regulation of natural gas and electricity prices has begun to slowly move away from the rigid system of pricing rules toward one in which market forces play a greater role. This transition can only go so far, however, as the presence of market power makes it undesirable to allow gas and electricity prices to move freely with market forces. It follows that government must continue to exercise its responsibility in assuring adequate emergency planning in the regulated sectors and for encouraging some adjustments during disruptions.

---

[6]Also, they are not motivated to invest in emergency preparedness measures, for example, unless the regulator specifically provides the incentive to do so. Regulators routinely allow for localized forms of emergency preparedness, such as reliability of the electricity system, but do not take into account the national interest in energy security.

Monopoly Power

Two kinds of market power are thought to exist in the international oil market: that exercised by the Organization of Petroleum Exporting Countries (OPEC) and oil companies (monopoly) and that exercised by importing countries (monopsony). The presence of either form of market power represents a market failure that could justify government intervention for security reasons. This section considers whether the exercise of monopoly power has been an important factor in explaining oil price behavior.[7] The following section deals with the monopsony issue.

Direct evidence of the exercise of monopoly power would occur if suppliers have acted to strategically withhold supplies from the market in order to raise the price. Indirect evidence would occur when oil prices failed to decline during periods of falling costs or falling demand. A brief look at oil prices will reveal surprisingly little evidence of the exercise of monopoly power, either by OPEC or by the oil companies supplying retail markets in the United States.

We start with a long-run perspective on oil prices. Figure 1 gives the constant dollar price of crude oil in the United States since 1860. Over the past hundred years, the price of crude oil has fluctuated within the surprisingly narrow range of $10 to $20 per barrel, except for the period from 1979 to 1985. This flat long-run trend does not convey the appearance of increasing scarcity one might expect with an exhaustible resource.[8] The flat trend also indicates that, to find evidence of market power, one should concentrate on the period from 1974 to 1985.

The price increases of 1974 and 1979 are widely regarded as the product of OPEC market power. However, the data do not support the argument that supplies were withheld from the market in these years to increase the price. As figure 2 shows, crude oil exports from OPEC countries and the rest of the world did not decline significantly in 1974 and actually increased in 1979.

---

[7]For different views on this issue, compare P. MacAvoy, *Crude Oil Prices as Determined by OPEC and Market Fundamentals* (Cambridge, Ballinger, 1982); and J. M. Griffen, "OPEC Behavior: A Test of Alternative Hypotheses," *American Economic Review* vol. 75, no. 5 (December, 1985), pp. 954–963.

[8]Production costs of exhaustible resources are in general expected to rise over time as the lowest-cost resources are depleted first.

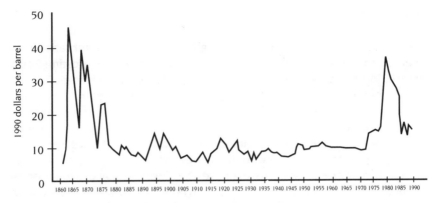

Figure 1. U.S. crude oil prices (converted to 1990 = 100).
Sources: U.S. Bureau of the Census, Historical Statistics of the United States, Colonial Times to 1970, Part 2, M138–142, pp. 593–594 (Washington, D.C.). Energy Information Administration, Petroleum Marketing Monthly, Domestic Crude Oil First Purchase Price, April 1990 and July 1990. U.S. Bureau of the Census, Wholesale Prices Indices, Producer Price Index.

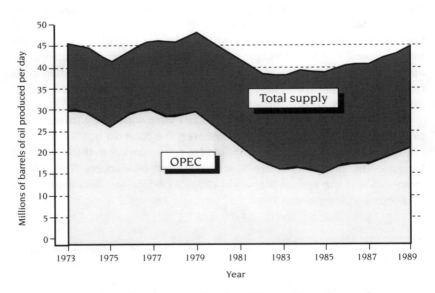

Figure 2. World crude oil production, in millions of barrels per day.
Source: Energy Information Administration, Monthly Energy Review, various issues.

An alternative explanation of the price increases can be found in actions taken on the demand side of the market.[9] The Arab oil embargo of 1974 and the Iranian revolution in 1979 did not significantly reduce oil supply, but these events apparently caused oil refiners, product distributors, and oil consumers to increase their stocks of crude oil and products. In effect, supplies were withdrawn from the market and added to inventories that would normally go to satisfy consumption requirements, causing a shortage far more important than that occurring on the supply side of the market.

After 1981, OPEC production shrank as consumption responded to high oil prices. However, OPEC could not slow production fast enough to prevent the price from weakening, and at the beginning of 1986 the price collapsed by 50 percent. If OPEC had possessed sufficient market power, this decline in the price would have been averted. Failure to stop the decline was not for lack of trying. OPEC repeatedly tried to discipline member countries to restrain output, but individual interests always dominated collective interests and the production quotas were not observed.

The data also give no indication that the domestic oil companies exercised control over product prices after 1973, as reflected in figure 3 by the close correspondence between the cost of oil imports and the price of gasoline in the United States. The price of gasoline follows a pattern just like that of the price of crude oil, rising in 1974 and 1979, trending down after 1981, and collapsing in 1986. There is no evidence, in other words, of an appreciable change in profit margins on petroleum products when feedstock costs were falling, as one would expect as indirect evidence of the exercise of market power.

Although there is no compelling evidence of the presence of monopoly power, there remains the fear that the OPEC cartel will eventually coalesce and control prices in the future. One quite ironic possibility is that the exercise of monopsony power by the United States and other importing countries could prompt OPEC to coalesce into an effective cartel.

---

[9]See D. R. Bohi, "What Causes Oil Price Shocks?" Resources for the Future Discussion Paper D-82S, January 1983 (Washington, D.C., Resources for the Future).

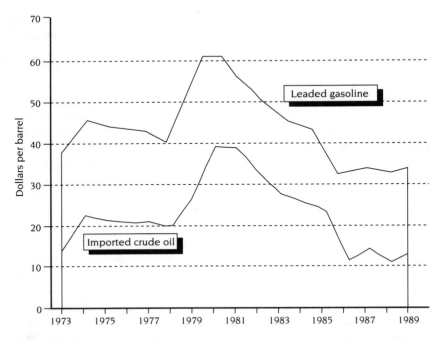

Figure 3. Gasoline and imported crude oil prices, in constant 1982–84 dollars.
Source: Energy Information Administration, *Monthly Energy Review*, various issues.

Monopsony Power

The United States may possess monopsony power over the world crude oil price because U.S. imports are such a large share of the total market. In 1990 the United States imported about 7 million barrels of oil per day, or about 12 percent of total world production, and 29 percent of internationally traded oil.[10]

Individual U.S. oil importers have no market power, however, and will not plan their purchases by taking into consideration their potential collective influence on the market price.[11] Their collective market power must be organized through government action. This mar-

---

[10]Energy Information Administration, *Monthly Energy Review*, (February 1991).

[11]The ability to influence the market price constitutes a market failure.

ket power could be used to force down the world price, which would benefit importing countries at the expense of exporting countries. It is a well-established notion in economics that there is an optimum tariff that would maximize the transfer of rents from producers to consumers.[12]

The foregoing argument is often used to demonstrate that the price of oil is too low and the volume of imports is too large.[13] There is merit in this argument, but also considerable controversy.[14] Three factors are especially critical in determining the conclusion. First, the tariff must result in a reduction in the world price or it creates no benefits for the United States.[15] There is, however, doubt about the effect of a marginal change in U.S. import demand on the world price. Part of the doubt is due to the lack of accurate estimates of supply and demand elasticities that are necessary to determine how the burden of the tariff will be split between consumers and producers.

The second consideration is the potential reaction of exporting countries, since the tariff will manipulate the world price to their disadvantage. Oil exporters could retaliate with trade restrictions against the United States, which would worsen the United States' position. Oil exporters could also restrict output in order to raise the world price of oil and negate the effect of the tariff. The latter risk is especially credible if the exporters have not been operating already as an effective cartel. If they are operating as a cartel, it is unlikely that further export restrictions are in the cartel's interest; if they are not operating as a cartel, a U.S. oil import tariff may be the event that galvanizes OPEC into effective action.

---

[12]See D. R. Bohi and W. D. Montgomery, *Oil Prices, Energy Security, and Import Policy* (Washington, D.C., Resources for the Future, 1982) for references to the optimum tariff literature and for applications of that literature to the oil market.

[13]See H. Broadman and W. W. Hogan, "Oil Tariff Policy in an Uncertain Market," EEPC Discussion Paper E-86–11 (Cambridge, Mass., Harvard University, 1986).

[14]See the survey of arguments in M. Toman, "The Economics of Energy Security: Theory, Evidence, Policy," ch. 25 in A. V. Kneese and V. L. Sweeney (eds.), *Handbook of Natural Resource and Energy Economics* vol. III (New York, Elsevier Science Publishing, 1993).

[15]There are distributional effects of a tariff to consider as well. Oil-producing regions of the country will benefit from a tariff, while oil-consuming regions will pay for these benefits. Also, individuals from different income classes may be affected differently by a tariff. An important factor in assessing the net distributional impact of a tariff is the disposition of tariff revenues.

The third consideration bears on an issue raised in the next section: does a tariff impose significant macroeconomic costs? This question is most critical in the midst of an oil disruption. If the economic costs imposed by a disruption are due to the effect of an energy price shock on macroeconomic performance (addressed in the next section), then a tariff raises the domestic price and hence the cost of a disruption even higher.

Macroeconomic Costs

Energy price shocks cause resources to be reallocated from one use to another, creating dislocations that can reduce total employment and output. For example, as energy becomes more expensive, the appropriate mix of energy and other inputs to use in production will change, and the desired mix of commodities in final consumption will also change. These adjustments reflect the inevitable reduction in the amount of goods and services that can be produced and consumed when energy becomes scarcer. The adjustment costs can be more serious, however, when institutional features of the economy interfere with the ability of the economy to adapt to the price shocks.

Wage rates, for example, tend to be adjusted infrequently (for example, once a year) and almost never in the downward direction. Nominal wages are therefore inflexible, especially in the downward direction, and the real cost of labor can be reduced only by unemploying workers or by inflating the economy. These conditions can present a problem when the price of energy suddenly rises, because a reduction in the use of energy will lower the productivity of labor.[16] Lower productivity means higher labor costs and increased pressure on employers to lower those costs. If the option of reducing wages is ruled out, employers are forced to reduce employment, which of course leads to further reductions in output.

Similar dislocations can occur because of other price rigidities throughout the economy, including the rigidities caused by regulation, as discussed earlier. The problem is essentially one of imperfect markets, where prices do not function to eliminate excess supplies and

---

[16]This assumes that energy and labor are complements in the production of other goods.

demands for resources and commodities immediately. Resources can become involuntarily unemployed (for example, since both workers and employers are committed to annual wage contracts) and commodity shortages and surpluses can develop. Eventually prices and wages would adjust to clear markets, but in the meantime an unnecessary and painful recession could get under way.

The foregoing argument has been used to explain why the oil price shocks of 1974 and 1979 caused recessions in the United States and many other industrial countries (see figure 4).[17] And, of course, we suffered another recession after Iraq's invasion of Kuwait. However, there are reasons to believe that the importance of these oil price shocks has been exaggerated. First, the current recession started several months before the increase in oil prices in August 1990. Second, the cost of oil is so small in relation to the size of the economy that it is hard to believe that even a doubling of the price of oil would have a measurable effect. For example, before the Iraqi invasion, the cost of imported oil amounted to about 0.5 percent of GNP. Thus, a doubling of the price of oil would reduce GNP by as much as 0.5 percent, assuming that none of the money spent on imports was respent in the United States. In contrast, over the past year GNP has fluctuated from quarter to quarter by an amount four times as large.

Third, if the oil price increases of 1974 and 1979 were so damaging, then for opposite reasons the price collapse in 1986 should have caused a worldwide economic boom. No such burst in activity occurred, however.

The final reason is brought on by asking why Japan did not suffer a recession in 1979, as did many other countries, even though Japan copied the other countries in 1974. One reason for the difference is that Japan did not impose a tight monetary policy on its economy in 1979, as it did in 1974, and as did all the other countries that suffered recessions in the two time periods. Deflationary monetary policies were popular during both periods in order to fight inflation, and rising oil prices were feared primarily because of their contribution to inflation. Consequently, output and employment were deliberately sacrificed as a mat-

---

[17]See M. Bruno and J. Sachs, *Economics of Worldwide Stagflation* (Cambridge, Mass., Harvard University Press, 1985).

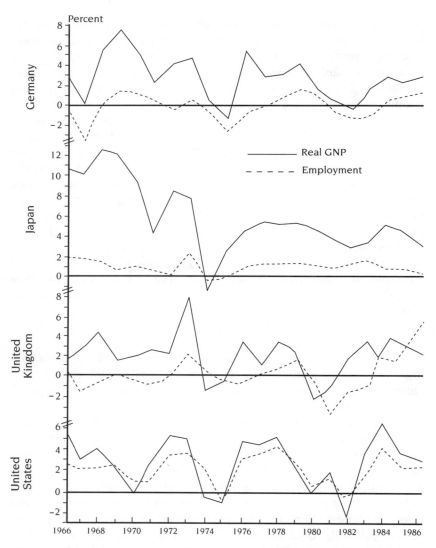

Figure 4. Annual percentage changes in real GNP and employment, Germany, Japan, United Kingdom, and United States, 1966–86.

Note: Real GDP is used for the United Kingdom.

Sources: Various issues of the International Monetary Fund's *International Financial Statistics* and of the Organization for Economic Co-operation and Development's *Labor Force Statistics* and *National Accounts*.

ter of macroeconomic stabilization policy. The recessions possibly could have been avoided if a little more inflation would have been tolerated.[18]

There are few policy tools to address these macroeconomic adjustment costs—since their source is the cause of the wage and price rigidities that slow adjustment—other than removal of regulatory interference wherever feasible in the economy. One option is to use monetary policy to inflate the economy and drive down the real cost of labor in order to remove the pressure to lay off workers. A second option is to blunt the oil price increase by releasing strategic oil reserves at the start of a disruption. Until 1990–91, neither of the two options had been taken by the United States.

Implications for Energy Security Policy

This discussion provokes surprisingly few clear recommendations about what the government should do for energy security reasons beyond what it is already doing. A case could be made for redirecting some R&D money toward diversifying our energy options, for example, but no unambiguous case could be made for a new policy initiative.

In view of this conclusion, a few words are in order about many of the policy options frequently suggested in the press, such as building a larger SPR, limiting oil imports, taxing gasoline consumption, subsidizing energy conservation, and subsidizing the production of alternative fuels. All of these options are often supported on the basis of some energy security externality.[19]

A larger SPR can be justified only by demonstrating that (a) it is an effective way to stabilize the world price and (b) price instability is costly for the economy. Doubts about the second condition were noted above. There is also reason to believe that the SPR, by itself, is incapable of exerting much influence on the world oil price in times of crisis. This follows because the maximum SPR draw-down rate (3.5 million

---

[18]For more on this topic, see D. R. Bohi, *Energy Price Shocks and Macroeconomic Performance* (Washington, D.C., Resources for the Future, 1989).

[19]These options may be justified on the basis of environmental externalities or achieving a balanced budget, but these issues are irrelevant here.

barrels per day) is small compared to the potential increases in inventories held by refiners, distributors, and consumers. Daily world oil consumption is 60 million barrels, but this rate can be augmented by several hundred million barrels of inventory demand in just a few months, easily swamping the contribution from the SPR.[20] This reasoning suggests that other importing countries would have to cooperate with the United States in order to exert control over the world price. The SPR is, of course, valuable in securing international cooperation, but the present magnitude of the SPR is sufficient for this purpose.

Limits on oil imports (through either a tariff or quota) can be justified on energy security grounds if, as noted above, the world price declines enough to make it worthwhile to burden domestic consumers and if exporting countries do not retaliate. Neither of these preconditions hold with reasonable assurance.

Oil import controls, gasoline taxes, and taxes on other petroleum products indirectly reduce the potential adjustment costs in the economy caused by an oil price shock, by forcing a reduction in the use of petroleum and reducing its importance in the economy. However, this argument amounts to imposing pain upon oneself in order to reduce the amount of pain that foreign disruptions can inflict. Self-inflicted pain may be milder, because one chooses the terms, but self-inflicted pain lasts indefinitely and oil market disruptions are temporary.

Finally, subsidies are recommended to stimulate the development of alternatives to fossil fuels, in order to achieve diversity in supply by improving the profitability of alternative fuels. Taxpayers who remember the defunct Synthetic Fuels Corporation will take hold of their wallets at this suggestion. Because government is not equipped to pick winners, government support for research is regarded as a more efficient way to achieve diversity than production subsidies. The same reasoning suggests that government support for research should aim at projects that are further away from commercialization; namely, for high-risk, basic research in which the private sector is less inclined to invest,

---

[20]Past behavior of inventory demand is based in large part on anecdotal information. Data are maintained on inventories held by refiners in Organisation for Economic Co-operation and Development countries. See Bohi, "What Causes Oil Price Shocks?," for a description of inventory behavior during the 1978–80 crisis period.

rather than for projects close to commercialization in which the private sector is most interested.

Much has been said in this paper about leaving matters to the private sector to deal with. It is worth emphasizing that the reason for doing so is not that markets are so perfect, but that the alternatives are so bad. One lesson from the energy crises of the 1970s that should not be lost is the damage to the economy that comes from the misguided actions of government policymakers. So far, in scoring market failures against government planning failures, the market wins.

Interdependencies Between Energy and Environmental Policies

W. DAVID MONTGOMERY

E nergy and environmental issues have been closely linked in the mind of the public, for good reason. A wide range of environmental problems are associated with energy production and use. Thus it is also presumed that environmental policy must have major effects on energy markets. But that, it turns out, is hardly true at all.

The processes of extracting, transforming, and using energy all have great potential for damaging the environment. The kinds of air pollution given the most attention over the last few decades in particular come in large part from the burning of fossil fuels. Nuclear power brings with it perceived risks of accidents and unanswered questions about the safekeeping of radioactive wastes. Production of oil and coal has created risks of oil spills and water contamination. And more recently all forms of fossil energy have been identified as contributing to the potential threat of global climate change.

Concern about the environment is also believed to have had major effects on energy markets. Fears about the environmental impacts of nuclear power are credited with bringing that once-

This paper was presented in the John M. Olin Distinguished Lectureship Series at the Colorado School of Mines on March 20, 1991.

promising option to a dead end. The potential for oil spills has led us to close off many promising areas to exploration or production. Energy conservation and the use of natural gas and certain forms of renewable energy have, on the other hand, been lauded because of what are believed to be relatively benign environmental consequences. Hundreds of millions of taxpayer dollars have been spent on developing "clean coal technologies."

Although the connection between energy use and environmental risks is unambiguous, the relation between environmental policy and energy markets is much less clear than this simple picture suggests. The types and amounts of energy that we use in the United States have actually been affected very little by the laws we have passed up until now to protect the environment. There have been major effects on how, where, and at what cost certain energy resources can be exploited—but not on how much we use. Oil imports may have been encouraged by restrictions on domestic oil production; production of coal has been stimulated in some regions and depressed in others. Costs of generating electricity, mining coal, and purchasing automobiles have been raised by the requirement to reduce emissions and effluents to levels that will cause no harm. Yet the overall amounts of oil, gas, coal, and electricity that we consume may have been changed only marginally by environmental laws and regulations. Even nuclear power may have met its fate for reasons other than environmentalists' opposition.

This decoupling of environmental policy and energy choices is due in part to market forces. Existing cost relationships among fuels and differentiation of their markets make it very hard to tip the balance away from one fuel and toward another. The design of environmental policy also plays a role. The Congress has in some cases consciously chosen to avoid or offset measures that would aid or harm one portion of the energy industry relative to another. In other cases a fundamental choice in U.S. environmental policy, to achieve emissions reductions through technology-based standards rather than economic incentives or behavior changes, has led to the decoupling.

The fact of decoupling has some interesting implications. The first is that some common ideas about ways in which energy policy could contribute to environmental improvement turn out to be plain wrong, and some ways in which existing energy policies are blamed for doing environmental damage are also quite wrong. The second is that if we changed some of the policy choices that lead to decoupling, we would see greater effects on energy markets but lower overall economic costs of achieving environmental objectives than under current policies.

Decoupling also has implications for business strategy. Market opportunities without question have been created and destroyed by environmental policy. Building scrubbers for coal generation plants will be a boom industry in the late 1990s owing to the new Clean Air Act; and building a brand-new, large energy facility, whether it be a greenfield oil refinery or a supertanker terminal, may be an option now foreclosed to U.S. companies. Nevertheless, an energy company that failed to pay great attention to environmental policy in making strategic decisions on how to allocate its resources among oil, gas, coal, renewable energy, and diversification outside of the energy field might have missed some interesting opportunities in the last decade, but it would not have been badly mistaken about the major forces driving energy markets.[1]

The new Clean Air Act Amendments continue this pattern. They will affect some market segments in ways that may lead to important business opportunities, but they will not make a large difference in the competitive balance among different fuels.

An issue on the horizon in the United States, but much closer in the rest of the world, may change all this. Concern about global warming, or more broadly, global climate change, focuses attention directly on fossil energy use as the culprit. Carbon dioxide emissions, which come not in traces or accidentally, but as one of the two compounds that fossil fuels become when burned, will have to be reduced to limit global warming. And the scale of the problem and the technological difficulty of removing carbon dioxide from waste streams after it is produced may overwhelm the competitive advantages among fuels and frustrate Congress's ability to diffuse impacts on specific energy sectors. Coal use must fall, relative to what it would be otherwise, and all

---

[1]The *Exxon Valdez* oil spill in Alaska's Prince William Sound cost the company $1 billion in fines and penalties. The CEO of Exxon noted that this settlement would have a "negligible impact" on Exxon's profits.

fossil energy use must eventually be reduced, if emissions are to be stabilized. But even in this case coal may prove, because of its cost advantages over natural gas, remarkably resilient in holding on to its share of existing electricity generation markets.

To foresee the outlook for energy markets if global warming becomes a major issue, it will be critical to understand how stringently and with what instruments policy to control carbon dioxide emissions will be designed. These potential changes would create new strategic opportunities, requiring decisions on when to abandon old lines of business and to move into new investments in forms of renewable energy and more energy-efficient products and processes. Rather than review every energy source, and every major environmental law or action from the passage of the National Environmental Policy Act to the present, I will illustrate how decoupling is accomplished and what it means by analyzing selected energy sources and recent environmental legislation.

My case, indeed, may be stronger in the area on which I will concentrate, the environmental and energy policies of the 1980s and 1990s, than in characterizing the effects of the environmental movement of the 1960s and 1970s. But even those days of activism may have had effects more localized than is normally realized.

Pollution and Energy

Let me begin by reviewing some of the major environmental issues in which energy plays a strong role. In air pollution the major issues throughout the 1980s were acid rain, reduction of urban air pollution in areas that are still in violation of the National Ambient Air Quality Standards, and control of air toxics. Energy is heavily implicated in the first two, acid rain and nonattainment areas. All coal contains some sulfur, and even with the levels of sulfur oxide removal required in new power plants by the Clean Air Act of 1974, emissions from coal-fired electric power plants in the Midwest are blamed as the primary cause of acid precipitation in the Eastern United States. Burning of residual fuel oil in Eastern power plants may also contribute.

Urban air pollution, typified by the smog of Denver and Los Angeles, comes from the release of hydrocarbons, nitrogen oxides, and carbon monoxide into the atmosphere. Motor vehicles, and the gasoline they burn, have been seen as the culprits. Tailpipe emissions of carbon monoxide, unburned hydrocarbons, and nitrogen oxides come from combustion of gasoline, and other hydrocarbon emissions come from evaporation of gasoline from vehicles and storage tanks and from filling up gas tanks.

Air toxics include a multitude—189 to be exact, for the Clean Air Act Amendments provide a list—of trace pollutants believed to pose acute dangers to human health. Such toxics are not so uniquely an energy phenomenon as acid rain and urban pollution. However, they are associated increasingly with another set of issues that relate to energy and environment, namely municipal solid waste combustion.

The Clean Air Act Amendments passed in 1990 mandate much more stringent controls on all these sources of pollution. Thus, I was astonished when I finally became convinced that the amendments are likely to do very little to change the amounts of coal and oil that we consume.

Two primary ways in which energy activities are implicated in water pollution are through oil spills and acid runoff from coal mining. The 1970 oil spill in the Santa Barbara channel and the 1989 *Exxon Valdez* incident focused public attention and outrage on the risks of oil production and transportation. These incidents, and others like them around the world, have helped to create a concern about the risks of oil exploration and development that has led the United States to close off many promising areas to oil exploration and development. Known large and productive oil fields, from which production is now banned, exist beneath the waters off California. The coastal portions of the Alaska National Wildlife Refuge also exhibit geological features that lend support to the speculation that a supergiant oil field could be found there. Allowing exploration of this area under strict environmental safeguards is one of the recommendations of the new National Energy Strategy, but whether this recommendation will prevail in the balance with perceived environmental risks is far from certain.

Coal mining is now a far less visible issue, but in the 1960s and 1970s pictures of the devastation wreaked by the acid runoff from coal mines created equal outrage. Worker safety and environmental health, including prevention of black lung disease, was also a major national issue. Important pieces of legislation, including the Mine Safety and

Health Act of 1969 and the Surface Mining Control and Reclamation Act of 1977, imposed strict new requirements on the coal industry, including prevention of runoff, restoration of mining sites to near-original condition, and dramatic changes in working conditions.

Finally, I come to the nuclear industry. From its promising beginning at Shippingport and Maine Yankee, this energy source has waxed and waned and is now on its way to at least temporary oblivion. Although the share of nuclear power in U.S. electricity generation continues to climb as nuclear units under construction are completed, no new nuclear generating unit has been ordered since 1976. Notice what that says about the unbelievable amount of time it has taken to complete the units that are still coming into service. Nuclear safety has always been a matter of environmental concern, even in the 1970s when the concern was largely theoretical and countered by reports of distinguished groups of scientists who claimed that major nuclear accidents were as close to impossible as any stochastic event could be (as, for instance, the 1975 reactor safety study commonly referred to as the "Rasmussen Report"). Three Mile Island and Chernobyl crystallized these fears, but long after U.S. electric utilities had stopped ordering nuclear units. Still unresolved is the issue of how to dispose of the radioactive waste that remains when uranium is depleted in generating electricity and when reactors must be disassembled.

The environmental heroes on the energy side are natural gas and certain forms of renewable energy, such as solar, wind, and certain forms of biomass energy utilization in which by-products are converted to useful forms of energy rather than being wasted or disposed of in environmentally damaging ways. Burning natural gas releases no sulfur and very little of any of the other pollutants that have mattered in the debates over the Clean Air Act. For many years the natural gas industry has hoped that environmental concern would open new markets and uses for its clean, safe fuel.

Although some forms of renewable energy, such as geothermal steam, have potentially severe pollution problems, solar and wind energy are seen as totally clean. Burning some types of biomass, such as municipal waste, can also contribute to air pollution unless carefully controlled, but reuse of "black liquor" as an energy source for the paper industry disposes of an extraordinarily damaging water pollutant harmlessly.

Market Review

Since market forces are responsible in large part for the decoupling of environmental policy and energy markets, let me review for a few minutes some of the characteristics of the structure and outlook for energy markets that matter most in this regard.

Oil. Oil is traded in a world market of which the United States has an important but not a dominant share. The United States imports a little less than half of the oil it consumes, a fraction that has been rising since world oil prices collapsed in the early 1980s. U.S. oil consumption is about 25 percent of the world's total, leaving aside those countries we used to refer to as Centrally Planned Economies (CPEs), on whose energy consumption we have little accurate data. As recent events have made us well aware, a large fraction of the world's oil supply comes from the Persian Gulf, although very little of that oil comes directly to the United States.

Oil markets have been on the verge of being glutted for some years, as the cartel of oil-producing countries, OPEC, found it difficult to agree on how and whether to restrain production to prop up prices. The Iranian revolution and then the invasion of Iran by Iraq (remember how we thought that was a good idea?) pushed oil prices to unprecedented levels in 1980. But by 1982 world oil markets became glutted as oil demand and production finally responded to high prices, and prices came tumbling down. The resulting tensions, which built within the cartel throughout the 1980s, may have added to Saddam Hussein's designs on Kuwait. The outlook for oil markets that prevailed before July 1990—and that now appears to be reconfirmed—was that oil prices would be stable at recent levels for some time, but that before the year 2000 world oil demand would have grown to such a degree that prices would have to rise. A midrange forecast might be for world oil prices to rise by about 2 percent per year between now and 2010. At these prices the United States can obtain imports freely to meet any increases in demand.

Natural Gas. Natural gas is a fuel whose market has been dominated by its own regulatory problems. Until 1978, natural gas prices were regulated by the Federal Power Commission, and wellhead prices were held well below market clearing levels. As a result, by the middle 1970s natural gas supplies fell short of demand, and natural gas pipelines were announcing curtailments, meaning that none but their highest-priority customers—such as residential customers—would receive supplies during peak winter months. What this meant was that, even if the Clean Air Act and public concerns about environmental consequences of oil and coal use had caused electric utilities or others to consider switching to natural gas, the supplies to allow them to do so were unavailable.

The beginning of partial deregulation under the Natural Gas Policy Act of 1978 (NGPA) allowed natural gas pipelines to bid for new gas supplies in order to serve electric utilities and other customers. With oil prices skyrocketing in the aftermath of the second oil shock, and because of some (intended) perversities of the NGPA, prices of some categories of natural gas were bid far above levels possible in a normal market. Pipelines also began to notice that what they had signed before deregulation had price escalators that became effective after NGPA raised wellhead price ceilings. The combination of rising prices of natural gas, driven up by NGPA and contract escalators, falling oil prices after 1981, and a recession drove natural gas demand down. The resulting oversupply of natural gas became known as the "bubble." For seven years now the ability of producers to deliver gas from existing fields at very low variable costs has held spot gas prices around $1.50 to $2.00 per million cubic feet, far below what most believe it would cost to replace those gas supplies once it is necessary to explore for and develop new reserves.

The outlook for natural gas prices is correspondingly uncertain. The conventional wisdom is that natural gas resources are limited in the United States, and that once the bubble disappears—as it must when existing fields are depleted—natural gas prices will be driven up by the rising cost of finding new supplies. This leads the Energy Information Administration to project that natural gas wellhead prices will reach $5.63 per million cubic feet in 1989 prices by 2010, at an escalation rate of constant real 5 percent growth per year. Yet we have also been predicting for the last seven years that the bubble will be gone in two to three years, and it has always remained that far off. The amount of natural gas found to be easily recoverable in and around existing fields has kept pace with extraction for all those years.

If it continues to do so, natural gas could remain a very cheap fuel for many years to come.

Coal Mining.   The environmental and health and safety legislation of the 1970s and later had a substantial impact on coal mining safety and costs. As a result, coal prices rose dramatically in the late 1960s and 1970s. Health and safety rules and acid drainage regulations may have forced the closing of some marginally profitable mines in the East and are credited with causing a substantial drop in labor productivity as expenditures on protecting employees grew.

Western strip mines whose environmental effects could be more cheaply controlled took up much of the slack. A major coal strike in 1973, coupled with rising oil prices that led utilities to increase their demand for coal, caused a 300 percent increase in coal spot prices in 1974. By 1975 coal prices stabilized at a level about twice that reached in the 1960s, but still only half the price of an equivalent amount of oil. The total demand for coal held up remarkably well through all of this. There was a shift in where and how coal was produced, from underground to surface mines, and from the East to the West, but the overall market for coal was not seriously affected. In part this may be due to the technological innovations that have by now reversed the decline in labor productivity. But it is much more due, as the Cost of Living Council concluded in a study of coal price escalation, to the segmentation of energy markets. The major market for coal is in baseload electricity generation and large industrial boilers. In this market it has a very substantial cost advantage over any alternative fuel. In other words, there were large rents in these markets. The costs imposed by these regulations were absorbed, by and large, in the rents, without altering coal's competitive position as the fuel of choice.

Market Segmentation.   The existence of such market segmentation does not affect coal alone. Oil has tremendous advantages as a transportation fuel, because of its high energy density and easy fueling and storage in vehicles. Natural gas has huge advantages over oil in other stationary applications, including industrial uses and electricity generation. Indeed, since the gas bubble appeared, it is generally accepted that natural gas has beaten out oil in every market it can enter. Electric utilities and industry only burn residual fuel oil in regions where there is inadequate pipeline capacity to deliver natural gas on a consistent basis. And natural gas has a strong advantage over coal in regions close to natural gas producing fields, in smaller industrial uses, and in electricity generation for intermediate and peaking loads.

Nuclear Power.   Nuclear power, as I mentioned, has had a history of rapid expansion and equally rapid decline. I think it was my colleague Jim Quirk at Caltech and I who called it a "rags to riches to rags" story, but I may be forgetting who we stole the line from. In the early 1970s the run-up in coal prices I mentioned earlier favored expansion of nuclear power for baseload generation. But as early as 1972 there were indications that the rate of escalation in nuclear power plant construction costs was beginning to tip the balance in favor of coal. By 1976 this trend was clearly established—estimates of nuclear power plant construction costs escalated by 300 percent between 1968 and 1976, while the general price level increased by "only" 67 percent. Orders for new nuclear power plants peaked in 1973, and since 1976 none have been placed. During 1973 another reason for the nuclear slowdown appeared—rising costs and inadequate revenues were making utilities unable or unwilling to finance any kind of capacity expansion. And shortly thereafter a third problem appeared: the unprecedented slowdown in demand for electricity that resulted from rising prices, recession, and mild weather—much like the problems facing natural gas in the 1980s.

Construction periods for nuclear power plants were long and grew longer. As a result, nuclear power plants still continue to come into service, and the share of nuclear power in electricity generation has grown from 9 percent in 1975 to 20 percent in 1990. But by 2010, under current policy, the share will only be 14 percent. However, policies that might change this trend, through resolving some of the sources of environmental concern, include identification of permanent waste disposal sites and movement toward modular construction, standardized design, and adoption of "inherently safe" reactor systems.

Electricity Generation.   Before leaving this discussion of energy markets, I want to mention a little about the outlook for electricity generation. Currently, coal accounts for about 55 percent of electricity generation, natural gas for 9 percent, oil for 5 percent, nuclear for 20 percent and hydropower for 11 percent. Coal and nuclear power plants have low operating costs, and are utilized as much as possible, while oil and natural gas-fired power plants are commonly cycled on and off to meet peak demands.

Although many electric utilities have had excess generating capacity for some time, that situation is rapidly ending. Utilities have a substantial amount of new coal-fired capacity under construction; as these large power plants also have substantial construction periods, those currently under construction will be completed between now and 2000. These planned additions will not be enough to meet all projected demand. For the most part, the additional units that utilities will build will be natural gas-fired units, some of them combined-cycle units that could be converted to burn coal at a later date. In this way natural gas could capture as much as a quarter of the market for new power plants over the next decade or two. At current prices the cost of generating electricity from a new natural gas-fired power plant is competitive in some regions with the cost of generating electricity from a new coal-fired power plant. In addition, natural gas plants are typically smaller, quicker to build, and easier to license, and have lower capital costs. Thus, when utilities are uncertain about how much electricity demand will really come about, and when they are financially constrained, they would prefer gas over coal even at some economic penalty. But if coal prices remain relatively flat while natural gas prices escalate as projected, coal will take over as the preferred choice in the latter part of the decade.

Environmental Policies

Let me now apply these observations about the nature of energy markets to the question of why it is that our major environmental policies have relatively little effect on energy markets. I will concentrate on the most recent of these policies, the Clean Air Act Amendments of 1990, and on current and upcoming issues. But to avoid the challenge that I am selecting the issues most supportive of my case, I will review a little of the past as well.

Oil Exploration and Development. Restrictions on oil exploration and development off the California coast may be credited to the fear of the repetition of the Santa Barbara oil spill. These restrictions and current prohibitions of exploration in the Alaska National Wildlife Refuge very likely make U.S. oil production, now and in the future, less than it would otherwise be. But we use no less oil because of this. Since oil imports are readily available, at the going price, on world markets, any oil not produced domestically is replaced by imports. There is, to be sure, some upward pressure on world oil prices because of higher import demand. And those higher prices will to some extent discourage domestic consumption of oil. But to the best of my ability to make the calculations, the increase in world oil prices caused by a loss of, say, 1 million barrels per day of domestic production would at most reduce oil demand by one-third of that amount, and likely by far less. So the effects of reduced production show up for the most part as increased imports and leave our use of oil nearly the same. There are also offsetting risks from these imports, as oil shipments to the United States on the high seas carry risks of oil spills no less—and possibly greater—than accompanies domestic production and transport of oil.

Nuclear Power. In the case of nuclear power, the question is whether the cost escalation that took nuclear power out of the running as an option for electric utilities was caused by environmental opposition or by other factors affecting the industry and its regulatory environment. Jim Quirk and I concluded, in a study we did during the 1970s,[2] that there were two important eras. Before 1970, intervenors imposed costs on nuclear power largely because of the procedural delays they were able to cause in licensing and construction. After 1970, the cost escalation largely came from the substantive changes in reactor design and construction required by the Nuclear Regulatory Commission. And some of the escalation came about because of mistakes made by electric utilities and the engineering and construction firms involved in building nuclear power plants. Some utilities were consistently able to build nuclear power plants within budget and on schedule, and others were continuously plagued with problems. The utility industry and its contractors seemed in some cases incapable of dealing with this highly

---

[2] W. David Montgomery and James P. Quirk, "Cost Escalation In Nuclear Power," in L. Ruedisili and M. Firebaugh, eds., *Perspectives on Energy: Issues, Ideas and Environmental Dilemmas,* edited (New York, Oxford University Press, 1978).

complex and demanding system. The existence of such management problems was confirmed by Three Mile Island.

Where this ends up in diagnosing the cause of the demise of the nuclear option I am not sure. Some responsibility must be assigned to the costs that were imposed by a regulatory process designed to provide ultimate protection to the environment and public safety, but some must also be assigned to the way the electric utility industry approached nuclear power. It may well be that in the case of both coal and nuclear power, regulatory processes did what substantive compliance costs could not: uncertainties about demand, delays in approval for siting, and low rates of return in the 1970s and 1980s turned utilities away from large, long-lead-time, capital-intensive generation toward gas-fired peaking and combined-cycle units. It may have been less the environmental impacts themselves, or the measures required to reduce those impacts, than the environmental, and more importantly the economic, regulatory environment in which electric utilities operated that slowed down coal and nuclear generation.

The Clean Air Act. Now we come to the environmental policies of the 1980s, and to their remarkable decoupling from energy market outcomes. The story begins with the Clean Air Act, which Congress tried to amend many times before finally succeeding in reaching agreement in 1990.

The impetus for passing the Clean Air Act Amendments came from the consensus that we were failing in three important areas—solving the problem of acid rain, achieving clean air in the worst-polluted urban areas, and eliminating risks to health from air toxics. What the Clean Air Act does is to impose much more stringent controls in each of these areas. In the case of acid rain and nonattainment areas, the pollution comes from burning coal and gasoline. Thus we might well expect that natural gas, or other clean fuels, would face greatly expanded markets.

The State of California has shown that there is a combination of market forces and policy designs that can achieve this result. Electricity generation in California is unique. No coal and almost no oil is burned in California for electricity generation. About half of electricity generation is from natural gas and half is from clean sources, including nuclear, geothermal, and (largely imported) hydropower. California's approach to coal has not been subtle: utilities may not burn coal within the state. There are big economic advantages in building coal-fired generating plants at the mine mouth in neighboring states and shipping the electricity to California, advantages that California has exploited. But even this accounts for only 2 percent or less of California's electricity. Natural gas is produced in California, avoiding high transportation costs, and hydropower is available. Thus, it appears that if natural gas and hydropower are available locally and in sufficient quantities, together with other energy sources, and if coal is directly banned, there can be an exception to the rule about decoupling.

But as it turns out, the exception may also prove the rule. National energy markets and environmental policy are very different from California's. And that is why the Clean Air Act Amendments will have surprisingly little effect on how much oil, natural gas, and coal we produce and use in the United States.

Decoupling comes in part from conscious policy design and in part from the segmentation of energy markets I described. The conscious policy has two parts, and goes back to the original Clean Air Act. These two parts are the reliance on technology-based standards and the determination to avoid damage to coal producers.

Our philosophy of air pollution control has always been to set standards for emissions from various sources based on the availability of control technologies that would remove pollutants from the waste streams. Implementing regulations specify the type of equipment that must be installed and kept operating or the maximum permissible emission rate allowed a particular source. Sources are defined very narrowly. A typical large manufacturing plant might have hundreds of sources, each with its own emissions limit.

An important regulatory innovation adopted in a limited way is the "bubble" concept, in which an industrial facility can depart from these individual source restrictions as long as its total emissions remain below the total of the emission limits set for all its sources. In effect, it can control some sources more tightly and some less tightly than the regulations require, as long as total emissions escaping from an imaginary bubble over the plant are not greater than they would be if each individual source were held to its permitted level. Trading of emissions permits between industrial facilities, one of which would exceed the standard and one of which would keep within it, has also been tried to a small extent.

But for the most part standards are written in terms such as "emissions from source x may not contain more than 1 pound of sulfur oxides per million Btu of energy input" or, in the case of motor vehicles, "1996 model year vehicles may not emit more than 0.25 grams per mile of hydrocarbons and 3.4 grams per mile of carbon monoxide." What this means is that changing the amount of time or the intensity with which a source is operated does not serve as a means of compliance, even if it would reduce the total amount of pollution being generated. The equipment required to keep the rate of emissions below the allowable level must always be installed.

Under the original Clean Air Act, electric utilities were placed under two types of control relating to sulfur oxides. Existing power plants were regulated by the states, under so-called State Implementation Plans, or SIPs, which in general did not require major modifications to the plants to reduce sulfur emissions. New plants were subject to New Source Performance Standards (NSPS), which essentially required that every new coal-fired power plant had to install a scrubber to remove sulfur oxides from stack gases after coal was burned. This requirement was stated deviously, in that the law required "uniform percentage removal" of sulfur from stack gases. The effect of the requirement was to eliminate switching from high-sulfur coal to low-sulfur coal as a means of compliance. This policy was adopted as a result of an alliance between environmentalists who wanted to require the maximum achievable reduction in SO_2 emissions, regardless of cost, and politicians who wanted to protect the interests of Eastern U.S. high-sulfur coal mines.

The NSPS requirement that all new coal power plants have scrubbers has increased the costs of building coal-fired power plants and may contribute to the share of natural gas in new power plants that I described in discussing the electricity outlook. But even if the cost of scrubbers adds 50 cents per million Btu to the cost of burning coal compared with natural gas, the advantage will disappear by the time natural gas prices reach the levels predicted for 2000. Moreover, just as the "uniform percentage removal" requirement removed switching to low-sulfur coal as a compliance option, so the NSPS in general provided little reason to reduce emissions by increasing reliance on existing gas-fired units and reducing utilization of existing coal-fired units.

The acid rain provisions of the Clean Air Act Amendments mandate much more stringent controls on sulfur dioxide emissions, but they also provide considerably greater flexibility in meeting these controls. Greater stringency comes through an emissions cap, which applies to existing as well as new power plants, while greater flexibility comes through the introduction of emissions trading and the fact that the cap applies to total emissions, not to the emissions rate. The emissions cap requires that sulfur dioxide emissions in 1995 be reduced to 2.5 pounds per million Btu times average fuel use in 1985–87. The cap will be ratcheted down in the year 2000 to 1.2 pounds per million Btu times 1985–87 fuel use. Because the cap applies to total emissions, emissions from new sources or increased use of existing sources must be offset. Each utility will be issued transferable SO_2 emission permits. Each annual ton of SO_2 emitted must be covered by an allowance, or the utility will be subject to a $2,000 per ton SO_2 fine. Allowances can be sold to another utility or banked for use in future years.

As long as an electric utility is expanding, it will be required to achieve an ever lower average SO_2 emission rate, probably approaching 1 pound per million Btu or less for some utilities by 2000. This is likely to require increasing use of scrubbers in existing units. But the change in the regulatory target, to tons of SO_2 for the entire utility instead of an emission rate (pounds per million Btu) for each plant, grants increased flexibility in fuel choice.

Compliance alternatives are considerably broader than they would have been under the standards typical of the old Clean Air Act. As before, utilities can install scrubbers on existing power plants in order to reduce emissions and make room, under their cap, for emissions from new units. But now utilities have the option of switching from high-sulfur to low-sulfur coal in existing units. Moreover, although new units are still subject to the NSPS requirement to have scrubbers, additional emissions reductions through use of low-sulfur coal in those units may also be valuable as a means of reducing the amount of offset required.

Reducing fuel input in the units with highest SO_2 emissions is also a compliance option. This could be accomplished through bulk power purchases from other utilities, by increasing use of gas-fired units, or by improving the heat rate of existing units. And for the first time, conservation actions that succeed in reducing SO_2 emissions become a compliance option under the Clean Air Act. Finally, emissions trading provides another mechanism by which compliance can be achieved. If buying an allowance from another utility is cheaper than any of the technological compliance measures, a utility can do so.

Interestingly, the Amendments do contain some reminders of the preference for technological controls; there is a "scrubber incentive" of additional allowances if scrubbers are installed prior to Phase II, and a clean-coal-technology incentive of delayed compliance and additional allowances for utilities that commit to adopting one of the specified clean coal-burning systems under development.

All of this suggests a potentially expanded role for natural gas and other clean fuels. The combination of increased stringency and flexibility should increase the opportunities and incentives for using natural gas instead of coal. The question is whether the combination is enough to overcome the inherent cost advantage of coal burned in existing power plants, or to make much difference in the economics of new power plants. Our analysis at Charles River Associates (CRA)[3] suggests that it is not enough. The cost advantage of coal over natural gas in existing power plants is so great that adding scrubbers or switching from high-sulfur to low-sulfur coal beats generating additional electricity from natural gas by a wide margin.

Some calculations we have done at CRA illustrate this analysis. Currently, low-sulfur coals are available at little premium over medium-sulfur coal. Thus, we calculate that a Midwestern utility switching from an average coal with 4 pounds of sulfur dioxide emissions per million Btu to a low-sulfur Midwestern or Powder River coal could reduce its emissions at a cost of about 3 cents per pound of sulfur dioxide removed. Installing scrubbers to remove sulfur dioxides at existing power plants would cost 13 to 18 cents per pound. But switching to natural gas at current average spot or contract prices would cost 28 to 36 cents per pound of sulfur removed. Although there may be some seasonal opportunities to use natural gas, both scrubbing and coal

---

[3]The author of this paper was Vice President of Charles River Associates when the lecture was delivered.

mixing appear to get most utilities under their likely system average target of 1 pound per million Btu at a much lower cost than natural gas, even at current prices.

In new power plants, as I mentioned, much of the added capacity to which utilities are not already committed is likely to be gas-fired between now and the year 2000. Utilities are not likely to cancel coal units already under construction in favor of additional gas capacity, for much the same reasons that they will not prefer gas in existing units. Paradoxically, the overwhelming success of natural gas in dominating the market for units to be built to meet peak demand growth for the next decade implies that there are no new fields for gas to conquer. Moreover, the ability of natural gas to compete with coal in new power plants has been based in part on the fact that many of these power plants are designed to meet peak demand, not continuous service in baseload, and in part on the current, abnormally low price of natural gas. If natural gas prices at electric utilities rise to the levels projected by the Energy Information Administration for the year 2000, nearly $4.00 per million cubic feet, coal will be far superior to natural gas in new power plants in most parts of the country.

Others have reached different conclusions. EPA appears to agree that the Clean Air Act Amendments will not reduce total coal consumption in electric utilities, but rather shift the mix from high- to low-sulfur coal. But EPA has also argued that the Amendments will cause some 1.2 trillion cubic feet (Tcf) of gas to be substituted for oil in existing dual-fuel power plants. I see little likelihood of this occurring because natural gas prices already favor gas over oil in these power plants. The only reason they would use oil is because of inadequate pipeline capacity to deliver gas when needed.

The Gas Research Institute (GRI) estimates an economic market potential for gas of 0.2 to 0.75 Tcf, with a technical maximum of 2.1 Tcf, due to the Amendments. They see substitution of natural gas for coal coming about through adoption of a process called gas/sorbent injection, which involves mixing gas and coal in modified boilers with another material that absorbs sulfur compounds. GRI also concludes that, although these processes are economic, adding scrubbers and switching to low-sulfur coal are cheaper for most applications.

The American Gas Association (AGA) has reached a more optimistic conclusion, that the options I have outlined could increase natural gas demand by 0.5 Tcf in 1995 and 1 Tcf in 2000, but their analysis assumes that existing price relations between coal and natural gas continue. If natural gas prices reach the levels projected by the Energy Information Administration by the year 2000, none of the AGA projections would make economic sense to utilities.

Thus there is a range of opinion on how much the Clean Air Act Amendments will affect natural gas and coal markets. But even if the AGA is correct, and gas demand is increased by 1 Tcf, gas sales would rise in the aggregate by only about 7 percent and still remain far below the peak levels of 1980. Coal use would decline by only 5 percent. It seems more likely to me that there will be no effect at all.

Where the Clean Air Act Amendments may make a difference is in the relative shares of high- and low-sulfur coal and in the fortunes of the regions in which they are produced. If utilities were not tied in to some supplies of high-sulfur coal by the long-term contracts that are prevalent in the industry, we would expect to see significant shifts to low-sulfur Western coal as the least expensive way to reduce emissions in existing plants and provide the allowances needed for new power plants. It may even be profitable for utilities to buy their way out of some of those contracts, in order to purchase low-sulfur coal and gain the economic benefit of the allowances using that coal will generate. As demand shifts to low-sulfur coal, especially that from the Powder River basin, its price may well be bid up. Now Powder River coal commands a price no higher than that of medium-sulfur midwestern coals, because until now the lack of incentives for use of any but the very cleanest coal ("compliance coal") made lack of sulfur an attribute with little value. Whether the price or the sales of Powder River coal will increase depends on how rapidly production or transportation costs will rise with increases in the scale of Powder River production.

The other parts of the Clean Air Act with potential impacts on energy markets are the clean-vehicle and clean-fuel provisions of the title dealing with nonattainment areas. Under these provisions, beginning in 1998 clean-fueled vehicles are to be introduced in about 25 urban areas with serious ozone or carbon monoxide pollution, including Denver. In California the program begins in 1996. These vehicles must meet standards more than twice as stringent as the new tailpipe emission standards that will apply to all vehicles by 1998. Manufacturers producing these vehicles will be allowed to use any combination of emission controls and clean fuels to meet the standards. Thus, the clean-vehicle program moves away from the technology-based standards of the old Clean Air Act, and also away from initial proposals in the administration's bill that would have required use of specific fuels, such as the perennial favorite, methanol. The major candidates for clean vehicles include methanol vehicles using a form of alcohol produced from natural gas, ethanol vehicles using alcohol from corn, ethanol vehicles using alcohol from corn, compressed natural gas (CNG) vehicles, and electric vehicles. A very low emission gasoline vehicle is not ruled out by the law either. Of these, methanol and compressed natural gas vehicles are now the most economic, for the uses contemplated.

The number of clean-fuel vehicles mandated by the Clean Air Act is not large. To bring these cars into use, once they are manufactured, operators of fleets of more than 10 vehicles, that are centrally fueled and not taken home at night or used in other ways that would make fueling difficult, are required to buy clean-fuel vehicles for a specified proportion of their fleet. The proportion rises from 30 percent initially to 70 percent in the third year. Interestingly, an allowance trading system for clean-fuel vehicles is to be set up, so that fleet operators with less than the mandated percentage of clean vehicles can come into compliance by purchasing allowances from fleet operators with greater than the mandated percentage.

The clean-fuel vehicle requirement is likely to result in some increase in natural gas demand at the expense of gasoline, because of the apparent cost-effectiveness of methanol or CNG for the kinds of fleet operations specified. However, since manufacturers are allowed to choose any combination of emissions controls and clean fuels to achieve the clean-vehicle standard, the possibility remains that reformulated gasoline will offer sufficient emissions reductions to be used. The Gas Research Institute has calculated that even if half of the fleet vehicles in the country converted to natural gas-based fuels, the increase in natural gas use would be about 0.2 Tcf, while the American Gas Association, which also includes in its calculations conversion of heavy trucks and transit buses to natural gas fuels, sees increases in natural gas demand as high as 1 Tcf.

Clean-fuel requirements are implemented separately from the clean-vehicle requirements. They apply to all fuels sold in the designated regions, during the times of year when pollution is worst. The requirements are designed to reduce both ozone and carbon monoxide in nonattainment areas. Reformulated gasoline is required in 9 ozone nonattainment areas, and oxygenated fuels are required in 40 carbon monoxide nonattainment areas (including Denver). Other states are allowed to "opt-in" to the clean-fuel requirements if they choose. The impact of the requirements on the cost of gasoline and on the refining industry hinges critically on how many do opt in. Refiners are required to meet these requirements for fuels sold in each nonattainment area, but again an allowance trading system is set up for each area so that fuel suppliers who fall short of the requirements can purchase allowances from suppliers whose products exceed requirements. Both the definition and specifications for clean fuels and the design of the trading system remain to be determined by EPA in upcoming rulemakings.

The way these fuels will be produced is by changing the composition of gasoline and including additives to increase the oxygen content. The cost of gasoline may increase significantly, by as much as 10 cents per gallon according to calculations we did at the Congressional Budget Office. Depending on how many states opt in, reformulated gasoline could account for between 25 and 50 percent of gasoline sales. These requirements could strain supply capability in the near term if many regions beyond those specified in the Clean Air Act opt in. Oxygenates could be a particular problem. Production capacity for one, MTBE, is limited, and would have to be used fully all year round to meet the seasonal requirements for oxygen content. Moreover, current supplies of ethanol, the other oxygenate, now used for gasohol in the Midwest would all have to be redirected to the problem regions. Since these additives are produced, respectively, from natural gas and corn, their use would reduce gasoline demand, but by less than 200,000 barrels per day.

If reformulated gasoline turns out to be as successful as some predict in reducing motor vehicle emissions, it might be the basis for a clean-fuel fleet based on gasoline, and eliminate even the fuel shifts indicated above for the clean vehicle program.

The Resource Conservation and Recovery Act. This act, known as RCRA, which governs the handling and disposal of hazardous and toxic wastes, comes up for reauthorization in this Congress. One of the esoteric issues in this law has to do with the classification of drilling muds, which are used in oil and gas drilling. These muds are now placed in a special category of high-volume, low-toxicity wastes and handled rather leniently, compared with other wastes covered by RCRA. There is some sentiment that this special classification should be terminated, and that drilling muds should be classified "Section C" wastes that require full RCRA toxic waste disposal licensing. Problems of regulating and permitting associated with this status could be as large as actual compliance costs. If this happens, handling of drilling muds would become much more difficult and costly, likely adding to the costs of domestic oil and gas production. But, as indicated in the case of restrictions on offshore oil development, the result would likely be some increase in oil imports, not reduction in oil use.

Energy Policies for the Environment

The decoupling that I have described as existing between environmental policy and energy outcomes also appears in the other direction. Some widely believed connections between energy policy and protection of the environment are also, in fact, nonexistent. One of the most popular misconceptions is that energy conservation must, of necessity, produce environmental gains. Although true in some cases, particularly if the conservation takes the form of reduced gasoline consumption through reduced driving, in other cases energy conservation does surprisingly little to benefit the environment. Reductions in electricity demand through energy conservation do almost nothing to reduce SO_2 emissions from electric utilities, and the Corporate Average Fuel-Economy (CAFE) standards, which reduce gasoline consumption in automobiles, may actually increase air pollution.

The reason that conservation of electricity does little for SO_2 emissions lies in the way electricity is generated. In the short run, a reduction in electricity demand leads electric utilities to cut back on the use of the power plants with the highest variable cost—which is made up almost entirely of fuel costs. In general, these are relatively clean gas- or oil-fired units, because the plants using dirtier fuels—coal and residual fuel oil—are the cheapest to operate.

In the longer term, utilities are likely to respond to reductions in electricity demand by eliminating the units whose retirement or cancellation will save the most money. This means, for the most part, canceling planned natural gas units, rather than shutting down existing coal units or canceling new coal units. As we just discussed, a disproportionately large share of new units to be added over the next decade will be gas fired, including almost all the units that could be canceled without penalty over the next few years. Planned coal units are mostly under construction already and subject to cancellation penalties. Thus, canceling new units means canceling a disproportionately large number of gas units. And running existing coal plants remains far less expensive than building a new plant of any type.

CAFE standards have for several years constrained motor vehicle manufacturers to increase the fuel economy of their fleets above levels that would prevail in the market. One effect of this improved fuel economy is to reduce the cost per mile of driving. And when driving is cheaper, the natural and well-established economic response is to drive more. Since vehicle emissions are not perfectly correlated with gasoline consumption, this increased driving could actually increase emissions over the level they would reach if market forces were allowed to determine fuel economy. Emissions are certainly greater than they would be if gasoline taxes, rather than technology-based equipment standards like CAFE, were used to promote energy conservation.

Another misconception is that federal subsidies to fossil energy—particularly oil and gas production—produce results that are harmful to the environment. Figures like $40 billion in federal subsidies to fossil energy are mentioned even in seemingly well-researched studies like the Oak Ridge National Laboratory's review of possibilities for energy conservation,[4] and blamed for encouraging wasteful use of fossil energy and tilting the playing field against more benign forms of energy. Yet almost all of these dollars provide subsidies that are inframarginal and do not even have an effect on production decisions or the market price of fossil fuels. Some are programs that have been terminated, like the many failed synthetic fuels programs, or that took the

[4]Roger Carlsmith, William Chandler, and coauthors, "Energy Efficiency: How Far Can We Go?" (Oak Ridge Tenn., Oak Ridge National Laboratory, January 1990).

form of investments in the development of technologies that are now fully deployed in the market, so that there is no longer an option to discontinue the subsidy. Others, like the much maligned oil depletion allowance, are designed to be inframarginal and have no effect on decisions to expand production. Although once available to all oil producers, the depletion allowance is now given only to the first 1,500 barrels per day produced by independent oil producers. The amount of oil eligible is very small, and increases in production by eligible producers above these limits do not benefit from the allowance at all. Tax preferences that do have an effect on the margin, like the investment tax credit, are available to all businesses and seem unlikely to redirect investment among lines of business.

Finally, even if federal budget dollars did stimulate domestic oil production, environmental damages would be very limited, for the same reasons that the environmental benefits of restrictions on offshore development are limited. Changes in domestic oil production are largely offset by changes in imports, leaving oil consumption and associated pollution unaltered.

Making Environmental Policy More Efficient

The decoupling of environmental policies and energy markets has come in part from policy choices that have avoided causing changes in patterns of energy use. For the most part, these choices have increased the cost of achieving environmental goals over what it would have been if changes in energy use had been allowed or encouraged. The most pervasive of these choices is the preference for technology-based standards over performance standards or the use of economic incentives to encourage behavior change as well as introduction of emissions control technology. The most egregious was the design of the original Clean Air Act to disallow switching to low-sulfur coal as a means of achieving compliance. This provision was consciously designed to prevent changes in coal demand, despite the substantial increases in compliance cost that it caused.

Emissions trading, such as that introduced in the Clean Air Act Amendments for sulfur oxides, changes the impact of environmental regulations on energy markets. In this case, trading, together with a cap on emissions expressed in total tons allowed rather than percentage removal required, causes significant shifts in the shares of Eastern high-sulfur and Western low-sulfur coal. But it has been credited with cutting in half the costs of keeping emissions below the cap, compared with what they could be if trading were not allowed. Trading provisions in the clean fuel and clean vehicle portions of the Amendments are actually likely to avoid disruptions of the gasoline refining industry that would occur if no flexibility were allowed.

A gasoline tax might be a far less costly means of reducing emissions that contribute to urban air pollution than many of the measures mandated by the Clean Air Act Amendments for serious and severe nonattainment areas. In order to bring these areas into compliance with the National Ambient Air Quality Standards, states are required to adopt control measures whose cost may range up to tens of thousands of dollars per ton of pollution removed. Even the measures that automobile manufacturers are required to implement on vehicles sold nationwide have costs in the thousands of dollars per ton removed. Yet some reductions in driving that would be brought about by a gasoline tax would reduce motor vehicle emissions at a far lower cost in lost consumer satisfaction. The lesson of all this is that abandoning attempts to prevent specific energy sectors from being adversely affected by environmental policy might serve to minimize economic damage overall.

Business Opportunities

The kinds of business opportunities and challenges created by the Clean Air Act have been interesting, even though their scope is limited in terms of the larger directions of energy markets. Emissions trading provisions of the act provide new opportunities for utilities to profit from choosing the most cost-effective combination of emission control measures and either offering excess allowances for sale or seeking out low-cost allowances rather than undertaking more costly scrubbing or fuel switching programs. Switching from high-sulfur coal to low-sulfur coal will also change the outlook for different segments and regions within the industry. But overall the Clean Air Act does not signal a great change in the fortunes of the coal or the natural gas industry.

Making the right moves to be in a position to produce clean fuels and vehicles could make a great difference to the fortunes of companies that have the technological capabilities to exploit the opportunities created by these portions of the Clean Air Act Amendments.

The Phase II reductions in SO_2 emissions required by the Clean Air Act will be difficult for many utility systems to meet unless scrubbers are installed on a large number of units. Building and installing scrubbers is a large and complex engineering and construction job, and the existing industry does not have the capability to do the required number of installations all at once. There is likely to be a real boom in demand for their services in the late 1990s. Since the financial penalties on a utility that is late in getting its scrubbers installed are severe, the bonding requirements on scrubber manufacturers and installers are likely to be large. The prices charged for scrubbers to be installed in 1999 are also likely to be high, with substantial discounts for earlier installation. Business opportunities abound here.

Despite these interesting possibilities, the Clean Air Act Amendments do not make the coal business look any less attractive than it did before they were passed. They will not return natural gas demand to the heady levels of a decade ago, and they will not create nationwide markets for new, clean fuels. Global warming may be different.

Global Warming

All around the world, concern has been growing that concentrations of greenhouse gases in the atmosphere are rising to levels that will cause global warming and widespread climate change. Among the most serious of the greenhouse gases is carbon dioxide. Energy is released from fossil fuels by combustion, which, as every chemistry student knows, turns them into carbon dioxide and water. Thus, carbon dioxide emissions are not an accidental consequence of energy-related activities, like oil spills. Nor are they the result of releasing traces of impurities in fuels, like sulfur, which can be cleaned up without interference with the basic processes of using energy. To keep carbon dioxide emissions down to levels that will avoid increasing concentrations of the gas in the atmosphere requires drastic reductions in fossil energy consumption. Thus, effective policies to combat global warming cannot be designed in ways that will preserve the decoupling from energy market outcomes that characterizes current environmental policy. Global warming is much the biggest environmental issue that has ever confronted energy markets.

Carbon dioxide is a particularly important greenhouse gas because once released into the atmosphere it stays there for many decades. Some natural processes, such as plant growth and dissolution in the oceans, remove carbon dioxide from the atmosphere. But the rates at which carbon dioxide is removed appear to be far less than the rates at which it is currently being generated. Thus, cumulative carbon dioxide emissions determine its concentration in the atmosphere. Rising concentrations trap greater and greater amounts of the sun's heat close to the earth's surface; this is the greenhouse effect that causes global warming. Although these general statements are agreed on by all, the questions of how much global warming is likely and what its consequences would be for global ecosystems and economies are as yet unanswered. Responsible scientists at one end of the spectrum recognize that the consequences could be so small as to be unrecognizable in comparison with all the other changes likely to be encountered over the next century. In some ways the changes could be beneficial. Scientists at the other extreme foresee catastrophic consequences, including widespread storms, inundations, desertification, famine, and extinction of many species. It will be decades before these uncertainties are narrowed. It is also likely that, unless the direst forecasts are realized, it will be 2050 or later before serious effects of global warming begin to be felt. These long time scales are fundamental considerations in designing global warming policies.

Because of this potential for disaster, many nations have announced their intention to stabilize carbon dioxide emissions at current levels or to reduce emissions significantly below current levels by 2005 or 2010. These nations include most of Western Europe and Japan. Although several bills were introduced in Congress in the last few years calling for stabilization or reduction of carbon dioxide emissions, the United States has not yet made any such commitment. It has been, however, an active participant in international panels studying the problem, and in meetings whose goal is to reach agreement on an international accord for control of greenhouse gases.

Given the policy positions of this administration, which I believe make a great deal of sense, action by the United States is likely to occur only in the context of an enforceable international agreement. Thus, one of the imponderables in predicting how global climate change policy will affect energy markets is in predicting when such policies will be implemented and what form they will take.

A widely discussed goal is that of achieving a reduction of carbon dioxide emissions to 20 percent below current levels by the year 2010 and holding emissions at or below that target for the remainder of the twenty-first century. I will first outline what policies designed to achieve this goal would mean for United States energy markets, and then turn to asking what an economically sensible policy approach would be. Then, reaffirming the economist's touching faith that good sense will prevail, I will discuss how business strategies might take into account the likely effects of sensible policies.

But before discussing what is sensible, let us ask first what would happen to energy markets if the United States adopted the goal of a 20 percent reduction below current levels by the year 2010. What we have talked about up until now should have established that how environmental policies are designed makes a great deal of difference in their impact on energy markets. Thus I need to specify how this goal would be achieved. I have studied the question by assuming that we would adopt the most cost-effective measures in order to meet the goal, and one way of accomplishing this is to use a carbon tax set at a level high enough to bring about the required changes in energy use.

It would appear, based on work we did at the Congressional Budget Office synthesizing the results of a number of studies,[5] that a charge rising gradually to $100 per ton of carbon in fuels by 2000, and settling at a level of $300 per ton in 2100, would be about right. The idea behind taxing carbon is that different fossil fuels release different amounts of carbon dioxide: coal is the worst, releasing about twice as much as natural gas, with oil falling right in the middle. The carbon tax provides an incentive to substitute away from the highest carbon fuels as well as an incentive to reduce fossil energy use overall.

---

[5]Congressional Budget Office, *Carbon Charges as a Response to Global Warming: The Effects of Taxing Fossil Fuels* (Washington, D.C., August 1990).

For reference, a carbon tax of $100 per ton is about $60 per ton of coal, or about three times the current minemouth price; $13 per barrel of oil, or maybe 75 percent of the current wellhead price; and $1.30 per million cubic feet of natural gas, again a bit less than the current wellhead price. In Btu-equivalent terms, a $100 per ton carbon tax is about $2.50 per million Btu of coal, $2.00 per million Btu of oil, and $1.30 per million Btu of natural gas.

Responses to a tax such as this will be cost-effective ways of reducing carbon dioxide emissions, and these responses will differ in the short, medium, and long term. In every time period there will be an effect on energy conservation, as the fuel taxes find their way into the prices of oil, gas, and coal consumed in the economy and into the prices of goods whose production uses energy as an input. Thus, energy conservation will occur as decisions are made to reduce the amount of energy consumed directly and as decisions are made to cut back on consumption of the nonenergy goods whose prices have been driven up most by the taxes and to substitute consumption of the goods whose prices have been driven up least. These can be summarized as changes in energy efficiency and changes in the structure of the economy. Energy conservation will be greater, all else being equal, the longer time there is for the stock of capital equipment and structures to be altered to the new desired patterns of energy use.

How, then, might fuel use change in the short, medium, and long term? Since coal is the most offensive fuel in terms of carbon dioxide emissions, we might again expect to see natural gas substituted for coal in electric utilities to achieve reductions in emissions. Different analysts have found different answers to this question, but once again, I find that coal exhibits a remarkable resilience and unwillingness to give in to gas.

One of the more interesting models of how energy markets and the economy will respond to a carbon tax has been developed by Dale Jorgensen and Peter Wilcoxen of Harvard.[6] They conclude that virtually all of the changes needed to reach the target of 20 percent reduction could be accomplished through electric utilities substituting natural gas for coal in electricity generation, and with a tax far lower than $100 per

---

[6]Dale Jorgensen and Peter Wilcoxen, "Reducing U.S. Carbon Dioxide Emissions: The Cost of Different Goals" (Cambridge. Mass., Harvard University, December 1990).

ton. Their model is at the forefront of econometric practice, and is in many ways the most detailed and theoretically appropriate model of the U.S. economy used for energy studies. But its representation of electric utilities leaves out some key factors. For utilities, as for all other sectors, Jorgensen and Wilcoxen assume that the capital stock is not specifically designed to use any particular fuel, that there are no technological constraints on substitution between fuels, and that there are no resource constraints on natural gas supply.

But technical and resource constraints, and the cost of converting existing utility power plants to other fuels, are the key to understanding competition between coal and natural gas. The carbon tax of $100 per ton adds about $1.30 per million Btu to the cost differential between coal and natural gas. This will certainly shift some dispatching decisions on the margin, toward gas use and away from coal. But the increase in the cost of all fuels will raise electricity prices and reduce demand. Thus, the initial impact of a carbon tax is mainly to reduce coal consumption in electric utilities, and natural gas consumption could also decline.

In a somewhat longer time frame, toward the years 2000 to 2010, when utilities can decide what types of new generating capacity to cancel, reductions in natural gas consumption are actually likely to be greater than reductions in coal consumption in electric utilities. Using a model that incorporates both the dispatching decisions and the decision of what type of new capacity to build, I find that by the year 2000 a carbon tax will reduce natural gas use in electric utilities by more than twice as much as it will reduce coal use, because so much of the new capacity to be added before 2000 is gas fired.

After 2000, electric utilities are likely to build a larger proportion of coal-fired power plants, if we believe forecasts like those of the Energy Information Administration in which natural gas prices rise relative to coal prices by more than $4.00 per million Btu. Thus, electric utilities have greater opportunities to compare the cost of new coal and gas units, to decide which to cancel. Since the carbon tax only raises coal prices by $1.30 per million Btu relative to gas, natural gas is at a greater disadvantage to coal in the post-2000 time frame than it is now, even after the carbon tax is taken into account. Thus, even by 2010, a carbon tax may cause greater reductions in gas consumption than in coal consumption in electric utilities. In all time periods, reductions in emission are tied to reductions in generation of electricity, not to fuel substitution.

But a carbon charge would not apply only to electric utilities. Since both transportation and other uses of petroleum products appear to respond strongly to price increases, there is likely to be a significant reduction in oil use in response to a tax that would amount to about 25 cents per gallon. Natural gas use would also fall in residential, commercial, and industrial sectors, and coal use would fall in the industrial sector. Many of these changes come about because the carbon tax gives an incentive not just for improving energy efficiency and switching fuels, but for changes in the structure of the economy as consumers move away from the consumption of energy-intensive goods. These changes in turn could have a dramatic effect on the fate of particular industries outside the energy sector, with increases and decreases in output as high as 10 percent of sales.

When we confine our attention to the next decade or two, the relatively modest increase in energy consumption likely to occur in the baseline makes achieving an emissions target 20 percent below current levels appear feasible. But as we look toward the middle of the next century, the picture is different. Not only will economic growth lead to growing energy demand, whose contribution to emissions must be fully offset, but the depletion of the natural gas resource base in the United States is likely to cause the share in total energy supplies of that clean fuel to dwindle away. Natural gas will no longer be available to substitute for coal or oil, and increases in energy consumption that come with economic growth will increasingly take the form of increasing coal use. The result of these trends is that by 2100 baseline carbon dioxide emissions could easily be four times their current level.[7]

By 2100 there will have been ample time to develop alternative sources of energy whose use will release no carbon dioxide into the atmosphere. Examples of these sources might be a new, inherently safe nuclear reactor that would be acceptable to the public and to electric utilities, or some form of renewable energy, such as solar energy or a form of biomass energy whose cultivation removes CO_2 from the atmo-

[7]Alan Manne and Richard Richels, "CO_2 Emissions Reduction: An Economic Cost Analysis for the United States," *The Energy Journal* vol. 11, no. 2 (April 1990).

sphere as rapidly as its combustion releases CO_2. A tax of $300 per ton has been estimated to be sufficient to overcome the cost advantages of fossil energy and lead to adoption of these technologies on a universal scale. Stabilization of emissions requires that all increases in energy demand be satisfied from a source that releases no emissions.

The difficulty will be getting through the interim period, say 2010 to 2040, when the commitment to stabilization must be met without the help of carbonfree energy sources available on a wide scale. The only remaining measure is raising taxes high enough to induce sufficient energy conservation to prevent growth in energy demand. That is a very costly operation.

Whether the reductions in fossil energy growth come from enforced conservation or substitution of renewable energy, the result is a much more bearish outlook for coal. In the baseline, coal forms the basis for most of the increase in energy use. Reducing emissions means drastic reductions from the baseline in coal use, probably to levels even below today's.

Policy Approaches

It can be seen from the numbers that I and others have presented that different fuels are conserved to different degrees in a cost-effective approach to reducing carbon dioxide emissions. As a result, sticking to the old policy approach of distributing the burden equally across all energy sectors could be very costly. In discussing my studies of carbon taxes with congressional staff, I heard numerous times the observation that "We hit the coal industry enough in the Clean Air Act. What about a Btu tax or CAFE standards as a way to reduce carbon dioxide?" A Btu tax, it turns out, is not so very different from a carbon tax in its effects, and so might not increase costs greatly. But attempting to shift the burden to a particular consuming sector and fuel (for example, the automobile), especially through an inherently inefficient regulatory system, could be very costly. The work of Jorgensen and Wilcoxen suggests that trying to achieve even modest emissions reductions through a gasoline tax rather than a carbon tax would multiply costs by a factor of 4. And a gasoline tax is clearly a more efficient instrument than some that might be chosen, like CAFE. Other efficiency standards, on buildings or industry for example, could also be very costly because they would defeat the shifts in the structure of economy that allow for some reductions in energy demand that are relatively cheap in the long run.

Another policy approach that makes a great deal of sense is buying information and developing new energy technologies before making the commitment to reduce emissions. The time scales for global warming are very long, affording time to do the scientific research required to narrow the range of uncertainty about what is likely to happen and what it will mean. Time is also available to develop new energy technologies that could be substituted for current energy sources at costs less than $300 per ton of carbon emissions avoided. Since, as far as the climate is concerned, it matters little whether we remove a ton of carbon now or in 2050, we can ask whether we should incur in 2010 the costs of drastic measures to reduce energy consumption below current levels, or invest in research and development that would allow us to achieve the same reduction at far lower cost in 2050.[8] The question is even more sharp when we realize that over that time period scientific research could convince us either that the problem is one truly requiring heroic measures or that the consequences of warming will be less costly to bear than the costs of avoidance.

Business Strategy

Just as global warming policy has the potential for much more profound effects on energy markets than environmental policies have had in the past, it has the potential for profound effects on the strategic outlook for energy businesses. There remain great uncertainties in this outlook, arising from the questions of whether, when, and in what form a commitment to emissions reduction will be made. It is necessary to weigh the likelihood of an immediate commitment to emissions reduction against the likelihood that governments will first commit resources to studying the problem and developing new energy technologies. Businesses that want to move aggressively must decide between acting in

---

[8]Steven Peck and Thomas Teisberg, "A Framework for Exploring Cost Effective CO_2 Control Paths," presented at a workshop on economic/energy/environmental modeling for climate policy analysis, Washington, D.C., October 22–23, 1990.

ways that will put them ahead of their competitors in capability to reduce emissions by 2010 and acting in ways that will put them ahead in developing the technologies for new kinds of energy supply and efficiency that will be used to address the problem effectively in much later years.

If an immediate commitment is made to stabilize carbon dioxide emissions, new markets will be created in equipment and processes that improve energy efficiency, from high mileage or clean fuel vehicles to new industrial processes. The outlook for coal will become bleak, and vigorous action will be needed to bring on line rapidly the currently available energy technologies from nuclear to biomass that could have zero net carbon dioxide emissions. In some cases it will be necessary to examine contending renewable technologies carefully, to see which are able to deliver on their promise of being carbon free. The problem with all biomass approaches is ensuring that additional carbon really is removed from the atmosphere, at appropriate rates, in providing their feedstocks. For example, harvesting standing timber for burning creates very large net increases in carbon dioxide emissions. Only rapidly growing crops, planted specifically as a source for biomass energy, can have zero net emissions.

If time is taken to ascertain how severe the problem is, even the dismal outlook for coal may be avoided, for there is always the possibility that the conclusion will be reached that affordable adaptation strategies dominate radical emissions reductions. Moreover, additional time can allow newer, more promising technological alternatives to be pursued, rather than committing to the deployment of the slim menu of alternative energy technologies available now.

Fortunately, taking time for scientific research and development on new technologies, accompanied by modest measures to promote energy conservation now, appears to be the policy of the Bush administration. I applaud them, and only wish that they would also admit the merits of using energy taxes as a mechanism for achieving cost-effective changes in energy use.

U.S. Energy and Environmental Policies: Problems of Federalism and Conflicting Goals

MARGARET A. WALLS

overnment intervention in energy markets ostensibly serves two purposes: improving energy security and correcting environmental problems. Both of these issues arise due to the existence of negative externalities in energy markets. These externalities drive a wedge between the private cost of energy—the cost to individuals—and the social cost of energy—the cost to the U.S. economy as a whole.

In correcting for these externalities, it is important to decide which level of government—federal, state, or local—is most appropriate for addressing the problem. On the one hand, our federalist system in the United States provides us the opportunity to reach more efficient outcomes than would a unitary system of government. Local and state governments can make decisions that are in their citizens' best interests without imposing significant costs on the country as a whole. On the other hand, the federalist system sometimes creates conflicts among

This paper was presented in the John M. Olin Distinguished Lectureship Series at the Colorado School of Mines on February 13, 1991.

The author greatly appreciates the research assistance of Carol Collins.

the jurisdictions. For example, a state could have a policy at odds with the objectives of the national government or vice versa.

In this paper I discuss the pros and cons of federal, state, and local intervention and look at the reasons for preferring one branch of government over another. In the section that follows, I introduce the basic rationale for government intervention, discussing all of the components of the energy security argument and the environmental protection argument. I then present a general introduction to federalism, and next show how current energy security and environmental responsibilities are assigned in the United States. Determinants of the optimal assignment are the focus of the next section. There I suggest that the optimal jurisdiction for provision of any public good is the one that is most capable of assessing and internalizing external costs and benefits; I then consider the circumstances peculiar to energy and the environment.

Moving from the world of economic theory to the real world, I look at two case studies of energy and environmental policymaking: federal government management of oil and gas resources on the Outer Continental Shelf (OCS) and federal regulation of motor vehicle emissions. Both of these studies are good examples of the potential problems of federalism and conflicting goals. I conclude the paper with broad prescriptions for determining optimal jurisdictions and ways of minimizing conflicts among jurisdictions.

Externalities in Energy Markets

In the market for imported oil, a wedge exists between marginal private and social costs that consists of two components: the *demand component* and the *disruption component*.[1] The demand component exists when the U.S. demand for imported oil is large enough to affect the world oil price. A simple example illustrates how this creates a wedge between private and social costs. Suppose that the United States imports 3 million barrels of oil per day for which it pays $15 per barrel. Suppose further that demand increases by 1 million barrels per day and that, as

---

[1]These terms are used by Bohi and Montgomery (1982), the best source for a detailed discussion of the energy security issue.

a result, the world price increases by $1 per barrel. The price of oil—that is, its marginal private cost—is $16 per barrel, but the marginal social cost is $19 per barrel ($16 × 4) − ($15 × 3). In this case, individual consumers in the United States do not take into account the impact that their own consumption has on the world price. A restriction on imports could be beneficial because the reduction in payments to foreign producers is more than offset by the value of oil to U.S. consumers. Figure 1 shows this graphically.

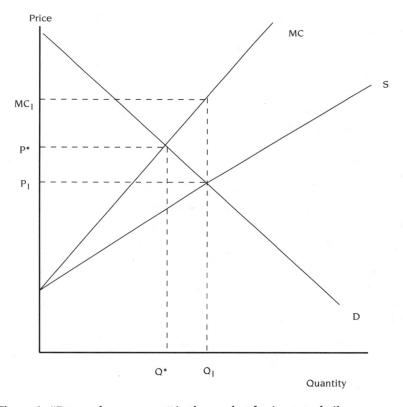

Figure 1. "Demand component" in the market for imported oil.
Note: D = import demand; S = import supply; MC = marginal cost of imports; Q_1, P_1 = private market equilibrium; MC_1 = marginal cost associated with Q_1; Q^*, P^* = socially efficient equilibrium.

The disruption component of the wedge exists when there is some probability of a disruption in world oil supplies that could significantly raise the price of oil. This future price increase will cause a transfer of wealth to foreign oil producers from U.S. consumers, just as described for the demand component above. In addition, the sudden price spike can cause dislocations to the macroeconomy. These come about when non-oil prices and wages cannot adjust immediately to rises in oil prices, causing various sectors of the economy to experience disequilibria.

Consumption and production of energy of all types, especially fossil fuels, also create environmental externalities. Production and refining of oil causes air and water pollution; mining of coal causes water pollution, soil erosion, and other disamenities; burning of fossil fuels in factories, electric utility generating plants, and cars and trucks creates air pollution. Once again, a wedge exists between marginal social cost and marginal private costs. No one individual producer or consumer has an incentive to take his or her contribution to pollution into account, even though the overall contribution has a clear negative impact on social welfare. This "free-rider" problem creates a role for government.

In many ways, improving energy security and reducing energy-related pollution are complementary goals. A policy that increases energy efficiency, for example, can have both energy security and environmental benefits. On the other hand, sometimes the two goals conflict. Limiting oil imports, for example, might decrease oil consumption and consumption-related pollution, but it would also increase domestic oil production and production-related pollution. Moreover, it might increase consumption of other fuels, such as coal, that are even dirtier. Likewise, policies to limit the use of coal might increase consumption of imported oil and reduce energy security. Even measures such as vehicle fuel economy standards, which are often thought to simultaneously reduce dependence on foreign oil and provide environmental improvements, may not be able to achieve both of these goals.[2]

---

[2]Although less gasoline is consumed per mile traveled, vehicles travel more miles because the cost of driving is reduced. Since motor vehicle emissions are defined for regulation purposes on a grams per mile rather than grams per gallon basis, total emissions can go up. For a discussion of the grams per mile versus grams per gallon debate, see Davis, Grusky, and Sioshansi (1989), and Khazzoom (1989).

Federalism

Overlaying these issues is the United States' federalist system of government. Although it may seem that most policymaking takes place at the national level in the United States today, this has not always been the case. Many historians and legal scholars view the period from the signing of the Constitution to the New Deal, and particularly the period up to 1861, as a period of "dual federalism"—a time when each level of government had its particular set of responsibilities independent of any other level.[3] The states had a great deal of autonomy during this period, and the role of the central government was minimal.

From 1861 to the New Deal, a gradual centralization occurred. During this period, two pieces of legislation with important implications for federalism were passed: the Interstate Commerce Act, prohibiting states from regulating industries in such a way as to impede commerce among the states; and the Sherman Antitrust Act. In addition, there was a move toward more cooperation and sharing of functions between the states and the national government. For example, this period marked the beginning of intergovernmental grants.

The New Deal initiated a virtual transformation of the U.S. political economy. Programs and regulation at the national level grew sharply, and intergovernmental relations and cooperation emerged as the new federalism.

Since energy and environmental concerns arose in the 1960s and 1970s, it is perhaps not surprising that the national government plays a large role in this area. By this period the national government was playing a large role in many areas. Passage of the major environmental laws—most importantly, the Clean Air Act (CAA) in 1970 and the Clean Water Act (CWA) in 1972—and the creation of both the Environmental Protection Agency (EPA) in 1970 and the U.S. Department of Energy's (DOE's) predecessor, the Federal Energy Administration, in 1974 are evidence of the importance of the national government.

---

[3]Mason (1972, p. 191) defines this system as the doctrine of "two mutually exclusive, reciprocally limited fields of power—that of the national government and that of the states. The two authorities confront each other as equals across a precise constitutional line, defining their respective jurisdictions."

On the other hand, both the CAA and the CWA, as well as other laws, leave room for state and local government participation. And the states and municipalities have embarked on programs of their own in the past three decades. Federalism—or "intergovernmental relations," as some legal scholars prefer to call the present system of government (see Reagan, 1972)—plays a large role in the making and carrying out of energy and environmental policies. The situation is a far cry from what it would be under a unitary system of government. Moreover, eight years of the Reagan administration, professing devotion to both deregulation and empowerment of the states, created a real role for federalist issues.

The Current Division of Responsibilities

In pursuit of energy security goals, the federal government stockpiles oil in the Strategic Petroleum Reserve (SPR) for release in the event of an import disruption. It leases federal lands to private companies for fossil fuel production. It provides tax incentives to domestic oil and gas producers (the depletion allowance and expensing of intangible drilling costs, for example). It sets fuel efficiency standards for cars and trucks, provides subsidies for production of alternative forms of energy (ethanol, for example), and supports research and development (R&D) of both conventional and alternative energy. One might argue, in fact, that the U.S. Department of Energy exists almost solely because of the energy security issue.

Many state governments mimic—though on a reduced scale—the federal policies. States lease lands for fossil fuel production and provide subsidies for R&D and production of alternative energy. In addition, most public utilities are required by their state public utility commissions to invest in fuel diversification and conservation.

Environmental regulation is also a mixed bag of federal and state, as well as local, intervention.[4] The federal government, through EPA, is responsible for setting standards under most major environ-

---

[4]See U.S. Congressional Budget Office (1988) for an overview of the allocation of environmental protection responsibilities between the federal and state governments. Also see Burtraw and Portney (1991) on this subject, and for a discussion of the major U.S. environmental laws.

mental statutes. The Clean Air Act sets national minimum ambient air quality standards for six major pollutants: particulates, sulfur dioxide, carbon monoxide, nitrogen dioxide, ozone, and lead. In addition, CAA sets national standards for motor vehicle emissions of hydrocarbons and nitrogen oxides (which combine to form ozone), carbon monoxide, and particulates. The Clean Water Act sets quantitative limits on effluent discharges into bodies of water. The responsibility for designing and implementing programs to meet these standards is primarily delegated to the states under both the CAA and CWA. Monitoring and enforcement responsibilities, including the levying and collecting of fines, invoking of civil and criminal penalties against violators, and revoking of permits, are principally delegated to the states. Research and development into the technical issues associated with standard setting, program design, and enforcement reside for the most part with EPA. DOE also carries out R&D for environmental purposes, the clean coal technology program being one major example. Funding for the environment comes from all levels of government, but many state activities are supported in part by grants from the federal government.

Local governments also undertake measures to reduce pollution. Sewage and water issues, of course, have always fallen to the local governments. But increasingly, municipalities are participating in other forms of environmental regulation. Perhaps the best example is the South Coast Air Quality Management District (SCAQMD), the authority responsible for establishing air pollution control programs in the Los Angeles area. Although states are responsible under the Clean Air Act for setting up State Implementation Plans (SIPs) for meeting national ambient air quality standards, California state officials have in recent years deferred to the SCAQMD to address the complex set of problems that are specific to Southern California. Other municipalities have also entered the environmental arena. Carpooling requirements, downtown parking restrictions, and mass transit are a few examples.[5]

---

[5]These measures also confer benefits in the form of reduced congestion, a goal that may be just as important to local governments as reduced pollution.

The Optimal Division of Responsibilities

The economist's notion of efficiency—the equating of marginal social benefits with marginal social costs in order to maximize the economic well-being of society—can be applied not only to determining the amount and types of policies government should undertake but also to deciding which level of government is most appropriate for those policies. In the broadest terms, that level of government should be chosen which (1) can most accurately assess external costs and benefits, be they environmental, energy security, or any other types of effects, and (2) is best able to internalize those external costs and benefits.

These criteria can be used to determine the appropriate jurisdiction for any publicly provided good or service. The optimal amount of police protection and trash collection in Denver, Colorado, for example, is more likely to be determined by the Denver government than by the Colorado or the U.S. government. The benefits of those services accrue almost entirely to the residents of Denver, and presumably the preferences of those residents are more apparent to the Denver government than to the government in Washington, D.C. It is interesting to note that this result of a government "close to the people" is consistent with the notions embodied in the Jeffersonian ideal of democracy.

Optimal provision of such things as defense against foreign aggression and a stable monetary regime, on the other hand, are more likely to be provided by the national government. The benefits of a defense provided by and for the citizens of Colorado, for example, would be likely to spill over into Wyoming. This free-riding phenomenon would lead to underprovision of defense for the country as a whole. Hamilton, Madison, and Jay may not have articulated terms such as "inefficiency" and "free-riding," but they clearly foresaw such problems when they advocated a strong national government, particularly in the area of defense.[6]

---

[6]See Jay's Paper No. 4 of *The Federalist Papers*, for example, in which he states that the national government "can apply the resources and power of the whole to the defense of any particular part, and that more easily and expeditiously than State governments or separate confederacies can possibly do, for want of concert and unity of system" (Hamilton, Jay, and Madison [1787], p. 48). Hamilton also showed some understanding of externalities in Paper No. 23 when he stated that leaving such items as national defense

Energy Security

We can use the above criteria to establish which level of government is most appropriate for achieving the goals of energy security and environmental protection.

The "demand component" of the energy security premium exists when the U.S. demand for oil is large enough to have an impact on the world oil price. As we explained above, individual consumers do not have the incentive to undertake socially optimal reductions in imports because individually they have no impact on the world price. By the same token, individual states cannot affect the world price either. As a result, policies instituted by the national government to limit imports may work, but state policies would not.[7] In this sense, the provision of energy security is like that of national defense. Because of free-riding, individual states acting alone would not provide enough of it.

The "disruption component" of the energy security premium exists when there is some probability of a future sharp reduction in oil imports and resulting rise in the world price of oil. The U.S. economy experiences both direct and indirect costs of such an outcome. The direct cost is the larger magnitude of the demand component during the disruption and, again, a policy to reduce imports is best instituted at the national, rather than state or local, level. The SPR exists to address such a problem. The indirect costs of the disruption are the dislocations to the U.S. macroeconomy that occur as a result of the inability of markets to adjust in the short run to a rapid increase in the price of oil. Release of oil from the SPR, by dampening the price increase, works to reduce these indirect costs, but a role for other policies may also exist.

The important difference to note is that the indirect costs depend not on the level of imports but on the level of total oil consumption in the economy. The United States could be self-sufficient in oil, but if its demand were large and price-inelastic, the economy could still feel sharp repercussions from a rise in the world price. Policies that help to

---

to the states would be "trust[ing] the great interests of the nation to hands which are disabled from managing them with vigor and success" (ibid., p. 155).

[7]The most obvious policy for limiting imports is a tariff, and of course only the national government is allowed to impose tariffs.

alleviate these indirect costs are ones that reduce the economy's over-all demand for oil—not just the demand for imports—and/or increase the elasticity of demand for oil. These goals can be accomplished by promoting the development of oil substitutes. This is where state or local government may have a role to play.

For various reasons, states may experience different levels of dislocation to their economies from an import disruption. The New England states, for example, have a large amount of oil consumption in the residential sector of their economies. This sector is likely to be less flexible in its demand for oil than, say, the industrial sector.[8] In addition, the burden of the price increase is not exported out of the region as it might be if the industrial sector were the predominant oil user. Indus-trial users may shift the burden to shareholders or to consumers of the final goods and services they produce, and these groups are located in all regions of the country, not just in New England.[9]

The New England states might prefer to invest in a larger amount of energy security than, say, Texas, where the largest oil consumers are industrial. A national policy would lead to underprovision of security for New England and overprovision for Texas.

The non-oil sectors of states' economies can also cause different amounts of economic dislocation. It is the existence of rigidities in non-oil prices and wages that causes the problems in the first place. If, for example, a large proportion of a state's labor force is unionized, one would expect less flexibility in wages in that state than in less unionized states and, hence, a larger disruption to its economy from an oil price rise. This state should reap more benefits from an energy security policy than nonunionized states.

Finally, states with large amounts of oil production experience positive impacts from a disruption and thus may prefer a small amount of energy security.

---

[8]See Bohi (1981) for a survey of energy demand elasticities in different sectors for different types of fuels. Most studies have found a larger (in absolute value) elasticity of demand for oil in the industrial sector than in the residential sector.

[9]Walls (1991), in a paper that estimates the regional impacts of an oil import fee, shows the breakdown of oil consumption and production by region, sector, shareholding in oil companies, and in the general corporate sector by region.

These examples indicate that states are more capable than the federal government of instituting policies that are tailor-made for their particular situations. The New England states may want to tax residential oil use or subsidize natural gas hookups in new houses or mandate insulation or appliance efficiency standards—any of a host of policies that would encourage flexibility in the residential sector. Highly unionized states may try to negotiate contingency plans with unions that would allow wage flexibility during an oil disruption. And states with a large amount of production might prefer severance taxes or windfall profits taxes that kick in when oil prices rise above a certain level. National instruments must necessarily be more blunt, since the administrative costs associated with finding the right policy for each of 50 states would be enormous.

There are some problems with leaving energy security provision to the states, however. First, the workings of states' economies are not isolated from one another. Auto manufacturers in Michigan use parts made in Kentucky and sell their cars in California. Michigan is unlikely, in the real world, to be able to smooth the adjustment of its economy to oil price shocks unless Kentucky and California cooperate with it. Moreover, Michigan could institute policies that appear to be in its own best interests but which would have negative repercussions in other states.

A second and related issue is that the shareholders of the auto manufacturers in Michigan are dispersed across the United States. A tax on oil consumption in Michigan might provide energy security benefits to the citizens of that state but is likely to be borne by the citizens of all 50 states, or at least those where the shareholders reside. In the long run, this inequity in the burden of costs and benefits would be likely to cause Michigan auto manufacturers to move to a different state, thwarting Michigan's efforts to improve its energy security situation. There are, of course, limits to this. If the automakers can pass some of the costs on to labor in the form of lower wages, for example, they would be less likely to relocate. A tax on residential oil consumption in New England, as I mentioned above, places the costs on the same individuals who experience the energy security benefits. Optimal provision of local public goods in fact requires this equating of the benefit recipients with those incurring the costs (see Musgrave and Musgrave, 1976, chapter 29). Such an equating is not always feasible in practice, however.

Third, the cost to industry of meeting 50 different requirements can sometimes be so high as to outweigh the benefits that the requirements generate. A useful example comes from the automobile industry. The auto industry experiences significant economies of scale in production. If a single state mandates higher fuel efficiency standards, the cost of automobiles sold in that state may be very high—higher perhaps than the energy security benefits provided by the reduced gasoline consumption. If other states imposed the same standards, however, the cost per automobile would be likely to fall significantly, perhaps enough eventually to make the regulation worthwhile.[10]

Finally, the information benefits of a policy instituted in a single state may not accrue just to that state. For example, a state program that helped to make photovoltaics economic, either through research and development expenditures or by actually subsidizing their use, might provide benefits to all 50 states. Individual states do not have an incentive to take these information spillovers into account and thus would provide a less than optimal amount of programs that generate such spillovers.

In conclusion, the argument for centralization of energy security policy is a strong one. The direct costs of the demand component and the disruption component must clearly be internalized on a national level. The macroeconomic dislocations of a disruption, while presenting some role for the states in promoting fuel flexibility and reduced oil consumption, also appear to need some federal government role because of the interdependence of states' economies, spillovers of either the benefits or costs of energy security, or economies of scale in the provision of energy security. The involvement of the federal government could take the form of either a supplanting of state regulations or supplemental grants to the states.

Environmental Protection

Although the role of the federal government in environmental policymaking is certainly significant, there are persuasive arguments for

---

[10]Interestingly, in a federalist system a single state that imposes a tighter-than-average regulation has an incentive to then lobby other states to pass similar regulations. By effecting such regulations, it lowers its own costs (see Rose-Ackerman, 1981).

decentralization. Clearly, the costs and benefits of environmental protection can vary widely from state to state and from city to city. For example, it is unlikely that Los Angeles would be able to comply with the national ambient ozone standard even if every car were removed from the highways there. Meteorological and topographical conditions play a large role in determining the cost of ozone controls in Los Angeles, as they do in other parts of the country as well. And there are other ways in which costs can vary across regions. Industries are often concentrated in particular regions of the country, for example, and it is more difficult for some industries than for others to cut back on pollution. Benefits can vary widely across and within states as well. One reason for this is variation in income levels. Environmental protection is a normal (maybe even superior) good, meaning that the demand for it rises as income rises. Moreover, population can matter. Improving air quality in a city brings much greater benefits than would the same improvement in a rural area. Air quality is a relatively "pure" public good; the more people "consuming" it, the more benefits it conveys.

In principle, it is possible for a national government to account for these differences in costs and benefits across regions and provide differing amounts of environmental protection as a result. In practice, the distance of the national government from the governed makes such an ideal outcome unlikely. The cost to the national government of determining all of the optimal amounts and then monitoring and enforcing the regulations would be large. A more likely result of national control—and the one that we see in the United States in most cases—is a set of uniform regulations across regions.

In addition to the greater ability of local governments to assess and respond to their citizens' needs, society can benefit from having a diversity of environmental protection levels. Because individuals have various preferences—for environmental quality as well as for other goods—they will choose the amount of the public good that is right for them. They can do this by moving to the community that provides the right amount.[11] Some communities might provide less environmental protection than neighboring communities but also tax their citizens less.

---

[11] This "voting with the feet" phenomenon was first identified by Tiebout (1956) in a classic article. It could apply to energy security as well.

We would expect to see individuals moving among the jurisdictions in response to their preferences and willingness to pay for environmental protection.

Some students of federalism find state and local governments to be more innovative and flexible in finding solutions to problems, including environmental problems, than the national government. Supreme Court Justice Louis Brandeis perhaps stated it best: "It is one of the happy incidents of the federal system that a single courageous State may, if its citizens choose, serve as a laboratory; and try novel social and economic experiments without risk to the rest of the country."[12]

Los Angeles again serves as a useful example. The 1989 SCAQMD plan for achieving the national ozone standard includes many "novel social and economic experiments," such as carpooling, flexible work schedule requirements, use of electric vehicles, and restrictions on driving schedules by area and time of day (see SCAQMD, 1988). These experiments may prove to be quite costly (see Portney et al., 1989), but nowhere near as costly as they would be if instituted nationally. By acting as a "laboratory," Los Angeles could confer positive externalities on the rest of the country; for instance, future costs of these programs may be lower as a result of their use in Los Angeles, and the "bad eggs" may be separated from the good ones.[13]

A last and more pragmatic reason for allocating responsibilities to the states is that this is often the de facto situation anyway. Obviously Los Angeles has not yet removed every car from its highways. In fact, most EPA officials, when pressed, would admit that they do not really expect Los Angeles to meet the national ozone standard. Certainly, they would not expect the city to go so far as to remove every car from its highways.

Despite this strong case for state and local decision making, there are a number of arguments for centralized control. Perhaps the most important one concerns the spillover of environmental problems across jurisdictions.

[12]*New State Ice Company v. Liebmann*. 285 U.S. 262, 311 (1932).

[13]Once again, however, Los Angeles may have an incentive to "lobby" the rest of the country to institute the same program if this would lower the program's cost (see footnote 10 above).

For years, the New England states have complained that acid rain in that region is a result of sulfur dioxide (SO_2) emissions from power plants in the Midwest. Clearly, contamination of the Chesapeake Bay comes from Delaware, Maryland, Pennsylvania, and Virginia; no one of these states alone is responsible. Such spillover of environmental problems means that individual jurisdictions acting alone will provide less than optimal amounts of environmental protection.

Benefits may spill over from one jurisdiction to another even when actual physical damages do not. For example, all U.S. citizens might be willing to pay some amount of money to keep the Grand Canyon pollution-free. This willingness to pay could arise because people place values on simply knowing that the Grand Canyon is pristine, even though they do not plan to visit it (these are called *existence values*) or because they want to preserve it for their own or their children's future use (these are called *option values* and *bequest values*, respectively).[14] The citizens of Arizona have no incentive to take these benefit externalities into account when establishing environmental regulations.

There are no constraints against cooperation and bargaining among the states to solve these types of problems—in fact, some cooperation has taken place with respect to the Chesapeake Bay—but, typically, coordinated efforts are difficult. The greater the number of states involved and the greater the distances between them, the more difficult those efforts become. Optimal policy may thus require some role for the national government.

Other arguments against decentralization are similar to those outlined above for energy security. First, it is important that the costs of tighter regulations are borne by those who reap the benefits. If one state imposes tighter-than-average environmental regulations, that state can expect to see its businesses relocating if the burden of those regulations is borne by the businesses' shareholders. The problem really is one of spillovers again, but this time they are cost spillovers. Some researchers have worried that this problem could lead to states competing for businesses by lowering their environmental regulations below optimal levels (see Cumberland, 1979). This destructive competition is said to lead to a role for the federal government. It is important

---

[14]See Mitchell and Carson (1989) for a discussion of these terms.

to note, however, that if the burden of the regulations can be shifted to workers in the form of lower wages instead of to shareholders, and if those workers cannot escape the burden by changing jobs, then the regulations would not cause businesses to relocate. To date, there has been little empirical study of this issue.

Another argument for a federal role is that there are benefits from uniformity. The costs to auto companies, for example, of meeting 50 different vehicle emissions standards would be extremely high. Many consumers would not be willing to pay these high costs for the environmental benefits received. On the other hand, because of economies of scale in production, the cost of meeting one standard might be low enough to generate positive net benefits.

A final argument often used to justify federal intervention is the presence of intergenerational impacts. Some current environmental damages—nuclear wastes and greenhouse gas emissions, for example—may lead to a loss of welfare for citizens in the next generation. Because there is a high probability that our children will not live in the same community that we live in, we do not necessarily have an incentive to force our local officials to take those benefits into account. On the other hand, we probably do believe that our children will live somewhere in the United States and we would like the national government to provide for future generations. Notice that this is just another kind of spillover, or externality. In this case, the benefits of environmental protection spill over into the next generation. The level of government that is best able to incorporate those benefits should decide on the appropriate level of environmental protection.

In conclusion, there is no one level of government that is appropriate for making environmental policy. To the extent that spillovers are minimal, the appropriate level of government is likely to be either state or local. These governments "close to the people" are more capable of estimating benefits and costs and are likely to design more flexible policies to deal with specific problems. If spillovers are important, the federal government should have a role. Good examples of such areas are global warming and acid rain. No individual nation, much less a state, has an incentive to act unilaterally to reduce emissions of greenhouse gases. The benefits of these reductions spill over to other states and nations, as well as to future generations. Acid rain is also an interstate, and often an international, problem.

The role of the federal government need not take the place of state and local controls. For example, the federal government could provide grants to the Midwestern states for the purpose of controlling SO$_2$ emissions. The amount of the grants could be based on the external benefits of these controls, that is, the benefits to New England from reduced acid rain.

These broad prescriptions for energy security and environmental problems can help us find efficient policies, but they will not eliminate all conflicts among jurisdictions. In the following case studies, I look at two areas in which the federal government intervenes ostensibly to serve energy security or environmental goals, and I consider the problems that arise due to federalism. In particular, I discuss the costs that states and localities bear for federal government policies that are seemingly in the national interest, suggesting general ways in which the situations could be improved. In the following sections, I first discuss the federal government's leasing of lands on the Outer Continental Shelf (OCS) for oil and gas production, and then its efforts to control emissions from motor vehicles.

Federalism Issues on the Outer Continental Shelf

The OCS is the area in the oceans that stretches from state water boundaries (3 miles offshore in most areas) to 200 miles offshore. It is owned by the federal government and leased to private corporations for the purpose of exploring for, developing, and producing oil and gas. The OCS Lands Act Amendments of 1978 state that policies and procedures should "expedite exploration and development on the OCS in order to achieve national economic and energy goals, assure national security, reduce dependence on foreign sources, and maintain a favorable balance of payments in world trade" (see U.S. Department of the Interior, Minerals Management Service, 1986, p. 6). Although environmental concerns are also addressed in the Amendments, this energy security goal is clearly the major one for the management of OCS resources. In the proposed final five-year leasing plan of 1987 the Materials Management Service (MMS) states that "it would be incongruous to restrict access to areas offshore which have high potential to contribute to our domestic production in a way that benefits the taxpayer and the economy in an environmentally safe manner, at a time when other options to increase energy security have substantial costs to the tax-payer and the economy" (U.S. Department of the Interior, Minerals Management Service, 1987, pp. 10–11). MMS further states that OCS production is a "necessary complement, not a substitute, for conserva-tion and alternative energy research efforts" (p. 11).

I argued above that the national government is the appropriate branch of government for addressing energy security concerns, although some state participation could be useful as well. In this sec-tion, I show that this prescription may be a naive one with respect to certain policies. The federal government's management of OCS oil and gas resources and its interactions with state and local governments have been fraught with problems.

Until the 1970s the coastal states had no say in the management of OCS oil and gas resources. The period from 1953 to 1972 was one of dual federalism in the management of ocean resources.[15] The states had control of resources in the territorial sea and the federal govern-ment had control of the OCS, and there was no overlap in responsi-bilities or cooperation between the two layers of government. Dual federalism ended with passage of the Coastal Zone Management Act.[16] The CZMA encourages coastal states to adopt comprehensive land use planning and management programs to guide coastal zone develop-ment. Participation in the program is voluntary, but states are induced to participate by two provisions in the act. First, they are given grants

[15]In 1953 the Submerged Lands Act turned the three-mile zone contiguous to the coast back over to the states. Ownership of this zone, known as the "territorial sea," had been disputed since the mid-nineteenth century. The Outer Continental Shelf Lands Act was also passed in 1953. This law gave the federal government absolute jurisdiction over the OCS and the authority it had previously lacked to issue leases to private companies. It also excluded the states from sharing in lease revenues. See Miller (1984) for a good discussion of these and other laws pertinent to the OCS.

[16]Although in principle the National Environmental Protection Act of 1969 gave state and local governments some say in federal activities, in practice the impact of this requirement has been minimal. The act requires the federal government to complete environmental impact statements before leasing, and to give citizens and government officials the opportunity to comment on these statements. Cicin-Sain and Knecht (1987) suggest that the process gives the "appearance of wide public involvement" (p. 164) but actually has little real effect.

to help pay for the programs. Second, once a state's plan is approved by the Secretary of Commerce, the state can require that federal actions affecting its coastal zone be consistent with its plan. It is this second provision that is important for states with OCS activities off their coasts. All federal licenses and permits that are required for OCS exploration, development, and production activities—and there are a number of them—must be consistent with coastal states' approved coastal zone management plans.

On the other hand, the CZMA also requires states to take account of national concerns in developing their coastal zone plans. They must give relevant federal agencies the opportunity to participate in development of the plans, and the Secretary of Commerce can deny program approval if he or she finds that the state failed to consider agency concerns. The 1976 Amendments to the CZMA required that states adequately plan for the siting of energy facilities and also created the Coastal Energy Impact Program, which provided grants to coastal states to mitigate the impacts of OCS development. This program was phased out in the early 1980s.

The OCS Lands Act Amendments of 1978 further provided for more state input into OCS activities. The Amendments require the Secretary of the Interior to propose a five-year leasing plan and submit the plan to the governors of the affected states for review and comment. Any comments that request modification require a written response from the Secretary. If the governors' recommendations are found to provide a reasonable balance between national interest and the well-being of the citizens of the affected states, those recommendations are used to revise the plan.

The Amendments also require lessees to submit exploration plans and development and production plans to the Secretary of the Interior prior to such activities. These plans must be approved by the governors and coastal zone management agencies of the affected states. If the plans are found to be inconsistent with coastal zone management plans, then the lessees can either modify the plans and resubmit them or appeal to the Secretary of Commerce.

The final provision of the 1978 Amendments that further considered states' interests was the establishment of the Offshore Oil Pollution Compensation Fund. This fund was created to provide for cleanup of and compensation for damages from oil spills and is funded by a 3 cent per barrel fee on oil produced on the OCS.

The OCS Lands Act Amendments of 1985 also slightly improved the position of coastal states. These Amendments require that 27 percent of receipts from OCS lands within a three-mile zone adjacent to state lands be shared with the coastal states. This sharing is designed to account for drainage of common pools (oil and gas reservoirs that underlie both federal and state lands). The Amendments also provided for distribution to states of funds from this zone that had been collected and held in an escrow account from September 1978 to October 1, 1985.

The final important federal law affecting coastal states is the Oil Pollution Act of 1990 (OPA). After the Exxon Valdez oil spill in March 1989, Congress scrambled to come up with an oil spill law that addressed spill liability, cleanup responsibility, and tanker regulations. The OPA increased the liability limits for tankers to $1,000 per gross ton under normal circumstances and to an unlimited amount if there was gross negligence or willful misconduct. If full compensation for damages cannot be obtained from the tanker owner, then additional compensation can come from a $1-billion federal fund created by the act. This fund is financed by a 5 cent per barrel tax on all domestically produced and imported oil.[17] Up to $1 billion may be spent on one spill, $500 million of which may go toward restoration of natural resources damaged by the spill. Importantly and controversially, the Act does not preempt stricter state liability laws.

The OPA gives the federal government primary responsibility for coordinating spill cleanup efforts. It also specifies that all new tankers be double-hulled and that existing single-hulled tankers be phased out by 1995.

Typically, the current OCS development process goes through the following sequence of events. The Secretary of the Interior releases the five-year leasing plan. State governors, local officials, and the general public, as well as other federal agencies and Congress, comment on the plan. Once the plan is adopted, MMS publishes in the *Federal Register*

---

[17]Money in the Offshore Oil Pollution Compensation Fund, created by the 1978 Amendments to the OCS Lands Act, was transferred to this new fund, and the 3 cent per barrel tax on offshore oil used to finance the old fund was eliminated.

a call for information and a notice of intent to prepare an environmental impact statement (EIS). A draft EIS is prepared and released to the public for comments, to be received within 60 days. Public hearings are also held within this 60-day period. A final EIS is then prepared, and a proposed notice of sale is announced by MMS. This notice is sent to the governors of affected states, who then have 60 days to submit comments on the size, timing, and location of the proposed sale. A final notice of sale is then prepared, taking the governors' concerns into account. The sale is held no less than 30 days after the final notice of sale. The process from adoption of the five-year plan to actual sale can take two to three years.

After a tract has been leased, the lessee submits an exploration plan to MMS. If MMS deems that all of the necessary information is contained in the plan, it is submitted to other federal agencies, the governor of the affected state, the state agency responsible for management of the coastal zone, and any other relevant state and local agencies. These agencies have a maximum of six months to determine if the plan is consistent with the state's coastal zone management program. If it is deemed inconsistent, the lessee must submit a new plan or appeal to the Secretary of Commerce.

Development and production plans follow much the same path as exploration plans. The only major difference is that often an EIS is required for development and production. This can delay the process by as much as two years. Any drilling, for exploratory or development purposes, requires a permit. These permits are granted by MMS and need not be approved by the state. The same is true of permits for platform construction. Pipeline approval and inspection also come from MMS, but if the pipeline goes through state waters, as it must to get to shore, then a copy of the pipeline application is sent to the affected state for approval.

The requirements embodied in the CZMA and the OCS Lands Act Amendments, the two main pieces of legislation that affect the OCS leasing and development process, have clearly led to a much greater role for state and local governments. They have also led to increased costs for OCS activities and added delays of many months—and in some cases years—to the development process. The main problem with the acts is that they have not addressed the issue that is really at the heart of the OCS debate: the coastal states bear all the costs of OCS activities and get very few of the benefits. Energy security benefits from reduced oil imports as a result of increased OCS production clearly go to the country as a whole. OCS revenues—lease bonus payments, rents, and royalties—all go into federal government coffers. States are prohibited by the Constitution from assessing severance or property taxes on federal properties. States with unitary corporate income tax structures can tax that part of OCS sales which fall within the state's jurisdiction, but this is not a large amount relative to total OCS revenues. The 27 percent of federal revenues from the three- to six-mile zone shared with the states is designed to compensate for drainage of common pools, not for any other costs such as the risk of an oil spill or the disamenities associated with oil and gas operations, be they aesthetic losses or reductions in recreation and tourism. And the one piece of legislation that did provide the states some compensation, the Coastal Energy Impact Program, was phased out in the early 1980s. The OPA provides for state compensation in the event of a spill, but the states get nothing unless a spill occurs, and then the payments simply make up for the damages created by the spill. Clearly, there are costs that the states incur even in the absence of a spill.

What the CZMA and OCS Lands Act Amendments have done is give coastal residents the means to delay, and in some cases halt altogether, the OCS development process. Coastal residents will attend public hearings, lobby government officials, and so forth, up to the point where the marginal benefit of that effort (the increased probability that the lease sale or development project will be halted) is just equal to the marginal cost (the value of other resources they have given up). It is perfectly rational behavior on the part of coastal states to attempt to delay OCS projects, when the cost of the projects are large relative to the benefits.

Of course, this divergence in costs and benefits varies across states. Californians place a high value on the non-oil amenities of their ocean and coastline. Moreover, their economy is a highly diversified one, less dependent on oil and gas than states like Louisiana and Texas. In fact, offshore oil workers and service and supply companies operating in California are often based in Louisiana or Texas—one other way in which the benefits are exported while the costs are borne locally. Out of approximately 153,000 jobs in Santa Barbara County, California, only 1,200 are in the oil and gas industry. Of the top 25 employers in the area, none is affiliated with oil and gas. By contrast, federal leases off Santa Barbara County generated, at the peak in 1984, $165 million in royalties alone (excluding bonus payments and rents).[18]

The Point Arguello project off the California coast north of Santa Barbara serves as a useful illustration of how long delays can sometimes become as a result of the stipulations set out in the CZMA and OCS Lands Act Amendments. The tracts that cover the Point Arguello field were leased in sales 48 and 53, in 1979 and 1981, respectively. The first discovery was made in the area by Chevron in 1981. Since that time, three platforms have been built, as well as an onshore processing facility, and offshore pipelines connecting the platforms to one another and to shore. All of the production and processing facilities were ready for start-up in 1987. At that point disputes over transportation of the oil arose. Chevron, the primary operator of the field, wanted to ship the oil by tanker to Los Angeles refineries. Santa Barbara County wanted the oil moved by pipeline.

In 1989 the county gave Chevron a permit for interim tankering. However, after the Exxon Valdez oil spill, the League of Women Voters and Get Oil Out (GOO), an environmental group, appealed the permit to the California Coastal Commission (CCC) and won. Chevron sued to overturn the decision and refiled the permit with the county, but the county sided with the CCC.

Since then—the county made its decision in November 1990—the debate has continued to rage. Chevron insists that the county's pipeline proposal is too costly, and the county insists that tankering, even temporary tankering while a special pipeline is built, is too risky.[19] Chevron offered to post a performance bond of $50 million to show its commitment to build a pipeline within four years, if the county would allow interim tankering. The county rejected this proposal.

---

[18]See Forecast '88/89 (1988).

[19]Santa Barbara County wants Chevron to transport the oil by an existing pipeline to a point near Bakersfield where a three-mile spur line would be built to another pipeline that goes to Los Angeles. Because Point Arguello crude oil is heavy, it would have to be blended with expensive light crudes that are in limited supply. Chevron claims that light crudes would have to come from outside the area by truck. They estimate the cost of this proposal at $2.40 per barrel (see Williams, 1990).

It is important to understand the magnitude of both the costs and volumes of oil associated with this project. The field has been estimated to contain between 300 million and 500 million barrels of crude oil, the largest oil field ever discovered on the OCS. At peak production, it is expected to produce 100,000 barrels per day. The cost of the development/production project alone was $2.5 billion, and Chevron claims that they spend $500,000 a day in interest costs (see Williams, 1990, and Tippee and Williams, 1990). Although the facilities were ready to begin full production in 1987, they are producing only a very small amount at the date of this writing.

The Point Arguello experience, along with a protracted period of low oil prices and two presidential administrations less supportive of OCS development,[20] is leading many oil companies to give up altogether on the California OCS. A "no development" policy may well be optimal for California. However, the process leading to that outcome is currently so costly and inefficient that it is difficult to judge. It is clear that the process wastes resources in administrative and litigation-related costs.[21] In the end, those costs contribute either to a cessation of the project—in which case the coastal citizens get what they want but the rest of the United States gets nothing—or they contribute stipulations that the project meet certain physical requirements—in which case the coastal citizens may get partial, but not full, compensation for putting up with the project and the rest of the United States gets more than it pays for. Unless some form of compensation or "side payments" for coastal residents can be instituted, efficiency losses are bound to result.

Federalism Issues in Regulation of Motor Vehicle Emissions

I argued above that environmental regulations should be established at the state or local level unless there are either spillovers of costs or

---

[20]In June 1990, President Bush declared a moratorium on leasing off the coasts of southwest Florida and most of California. President Clinton also has promised to restrict leasing.

[21]Chevron spent $3 million just on the EIS for the Point Arguello project (see Cicin-Sain and Knecht, 1987).

benefits to other regions or generations or there are economies of scale in production that outweigh the benefits of localized control. Pollution from automobiles does not spill over to any great extent from one state to another except in the Northeast. However, automobile production does exhibit significant economies of scale: the cost per vehicle of meeting one uniform regulation is much lower than the cost of meeting 25 or 50 different ones.

The latter point has dominated U.S. policymaking toward the automobile. Emissions standards are set at the national level and, except for California, preempt state standards.[22] As I will discuss below, this policy imposes high costs on some regions of the country that receive very few benefits. In this respect, the federal policy on emissions is like the federal OCS policy, although unlike the case with the OCS, state and local governments cannot do battle with the federal government over motor vehicle emissions, that is, there is no "consistency" provision as there is in the Coastal Zone Management Act that forces the federal government's actions to be consistent with the state's wishes.

Southern California began worrying about motor vehicle emissions in the late 1940s and early 1950s when it became clear that such emissions were a major component of urban smog. In 1963 California enacted a law that required exhaust emission control systems to be installed on all new cars sold in the state. The 1966 model year cars were equipped with these devices.

The California action spurred Pennsylvania and New York to push for their own regulations, and New York's proposed requirements were even more stringent than California's. These moves prompted the automobile industry to lobby Congress for national standards. The industry ended up lobbying for something that was seemingly not in its best interest because it did not want to incur the costs of meeting different standards in different states. Elliott, Ackerman, and Millian (1985) also claim that the industry preferred arguing technological issues with a federal agency over arguing symbolism with state politicians.[23] In 1965 Congress passed the Motor Vehicle Air Pollution Con-

---

[22]National standards are the norm for fuel efficiency and safety as well.

[23]The chemical industry finds itself in the same position today. To avoid meeting many different state requirements, it is pressuring Congress for national standards (see Kriz, 1989).

trol Act and amended the Clean Air Act. The latter directed the Secretary of Health, Education, and Welfare (HEW) to set national auto emissions standards. The 1968 model year cars were the first to meet national standards.

The Clean Air Act Amendments of 1970 marked a new era in pollution control, one of increased centralization. The 1970 Amendments were partly a result of increasing public concern about the environment in general and partly a result of frustration with HEW's dilatoriness in studying health effects of certain pollutants and the auto industry's slowness in lowering vehicle emissions (see Portney, 1990). In an unusual move, Congress wrote the actual standards into the act itself. The standards called for a 90 percent reduction in average hydrocarbon (HC) and carbon monoxide (CO) exhaust emissions and an 82 percent reduction in nitrogen oxide (NO_x) emissions over 1970 levels by 1975. These percentages translated into exhaust standards of 0.41 grams per mile (g/mi) HC, 3.4 g/mi CO, and 0.4 g/mi NO_x.

The 1975 deadline was eventually pushed back to 1978, but the auto industry was not even able to meet that deadline. As a result, Congress passed Amendments to the Clean Air Act in 1977. These Amendments delayed achievement of the HC standard to 1980 and the CO standard to 1981; the NO_x standard was loosened to 1.0 g/mi and delayed until 1981. No changes were made to federal auto emissions standards in the 1980s; the next major piece of legislation on this issue was the Clean Air Act Amendments of 1990 (CAAA).

The mobile sources provisions of the CAAA were a hotly contested feature of the bill. President Bush's original plan called for a tightening of standards in all areas and mandated the use of alternative fuels in the nine U.S. cities with the worst ozone problems. This sparked a flurry of lobbying for and against alternative fuels by the oil industry, the auto industry, and environmental groups. The end result was a mix of technological fixes to automobiles, alternative fuel requirements, requirements for changing the composition of gasoline, and a tightening of exhaust emissions standards. By the 1996 model year, all light-duty vehicles must meet a 0.25 g/mi nonmethane hydrocarbon (NMHC) standard, a 3.4 g/mi CO standard, and a 0.4 g/mi NO_x standard. The evolution of federal automobile emissions standards is shown in table 1.

The CAAA also requires that EPA carry out a study of whether standards twice as stringent as those above should be required after

Table 1. Federal Automobile Exhaust Emissions Standards
(in grams per mile)

Model year	HC	CO	NO$_x$
Precontrol	10.60	84.0	4.1
1968	6.30	51.0	(6.0)[a]
1970	4.10	34.0	—
1971	4.10	34.0	—
1972	3.00	28.0	—
1973	3.00	28.0	3.0
1974	3.00	28.0	3.0
1975	1.50	15.0	3.1
1976	1.50	15.0	3.1
1977	1.50	15.0	2.0
1978	1.50	15.0	2.0
1979	1.50	15.0	2.0
1980	0.41	7.0	2.0
1981	0.41	3.4	1.0
1982–1995[b]	0.41	3.4	1.0
1996–2003	0.25	3.4	0.4
2004[c]	0.125	1.7	0.2

Sources: Anderson (1990) and Clean Air Act Amendments of 1990.

[a]NO$_x$emissions were uncontrolled and increased with the declines in HC and CO.

[b]40 percent of 1994 model year cars and 80 percent of 1995 model year cars must meet the 1996 standard.

[c]These standards apply if it is found that the 1996–2003 standards are not expected to bring areas into attainment with national air quality standards.

2003 in order to meet ambient air quality standards. If the study determines that they should, the standards are to be promulgated. EPA is required to set standards for evaporative emissions within 18 months after passage of the CAAA. Also within 18 months, EPA must complete a study of the effects of toxic substances present in gasoline, such as benzene, formaldehyde, and 1,3 butadiene, and promulgate regulations for such substances within 54 months of passage of the CAAA.

The Amendments also require manufacturers to install on-board diagnostic systems on all new light-duty vehicles within 18 months.[24] Stage II vapor recovery systems are required on all gasoline pumps in certain ozone nonattainment areas, and the Reid vapor pressure (RVP)

---

[24]These systems identify deterioration or malfunction of emissions-related systems; they include catalytic converter sensors and oxygen sensors.

of gasoline sold during the summer months in these same areas must not exceed 9.0 pounds per square inch (psi).[25] Gasoline sold in certain ozone nonattainment areas must be reformulated to meet the following specifications: a benzene content of not more than 1.0 percent by volume; a total aromatic hydrocarbon content of not more than 25 percent by volume; no lead; additives to prevent deposit accumulation in engines; and an oxygen content of at least 2.0 percent by weight.[26] If a refiner or marketer sells a gasoline that more than meets the above requirements, he or she can obtain a credit that can be traded with other marketers in the same area. Credits cannot be traded with marketers in attainment areas if they lead to higher aromatics or lower oxygen content of fuels in those areas. Areas that do not achieve the CO standard must sell gasoline with a 2.7 percent oxygen content during the winter months. Credits also apply here and can only be traded within or between CO nonattainment areas. A tighter requirement of 3.1 percent oxygen will apply if the CO standard is not met.

The CAAA is the first law to include alternative fuel requirements. All fleet vehicles in selected ozone nonattainment areas must run on a "clean" fuel and meet stricter standards than those for conventional gasoline vehicles. The clean fuel standard for NMHC is half of the conventional fuel standard for 1996 and another 40 percent lower for 2001. The NO_x standard is half of the conventional fuel standard for 2001, and the CO standard is unchanged. A clean fuel can be methanol, ethanol, reformulated gasoline, diesel, natural gas, liquefied petroleum gas (propane), hydrogen, or electricity. All fleets are covered by the law except those for rental to the general public, law enforcement, and a few other categories. Flexible-fuel or dual-fuel vehicles, that is, vehicles that can operate on either a clean fuel or conventional gasoline, must meet slightly less stringent standards. Credit trading similar to that outlined above for reformulated gasoline is allowed under this program as well.

The CAAA also establishes a California pilot program for clean-fuel vehicles. Under this program, 150,000 clean-fuel vehicles must be

[25]Ethanol fuel receives a waiver of this requirement.

[26]Reformulated gasoline can meet either these requirements or requirements that HCs and toxics emissions be certain percentages below those from conventional gasoline.

sold to the general public in 1996, 1997, and 1998, and 300,000 must be sold per year thereafter. A credit program again applies. One last interesting feature of the CAAA is that other states may choose to institute California's tougher emissions standards. California has always been allowed to have its own set of standards, not preempted by federal regulations, but no other state has this prerogative. Other states may also "opt-in" to the California alternative fuels pilot program.

The success of the 1970s regulations in lowering vehicle emissions is well established. White (1982) has shown that, controlling for vehicle characteristics and mileage, HC and CO emissions from a 1979 model year car are roughly 30 percent, and NO_x emissions half, of what they would have been in the absence of controls.

The real question, of course, is what impacts the regulations have had on individual well-being. Estimating these impacts requires tracing changes in emissions to changes in ambient air quality, changes in air quality to changes in health and welfare, and, finally, changes in health and welfare to dollar estimates of benefits.

Carbon monoxide in high concentrations tends to replace oxygen in the bloodstream; this can lead to impairment of the nervous system or heart functions and is particularly problematic in people who suffer from angina. HC and NO_x combine to form ozone. Ozone can cause coughing and chest discomfort and is particularly bothersome for asthmatics. Ozone also can lower agricultural output and reduce the growth of trees and vegetation; it also reduces visibility, since it is the main component of smog. Average ambient concentrations of CO and ozone declined significantly over the 1972–87 period (see White, 1982, and Portney, 1990). The CO drop has been particularly dramatic: despite increases in both the number of vehicles on the road and the average number of miles traveled per vehicle, annual ambient CO concentrations fell by 32 percent between 1978 and 1987. Ozone has been a tougher problem since the sources of volatile organic compound (VOC) emissions are ubiquitous.[27] Nonetheless, ozone concentrations over the same period fell by 9 percent.

---

[27] VOCs are a set of compounds that include HCs, among other things. Motor vehicle VOC emissions are almost entirely composed of HCs, so the terms can be used interchangeably here.

Despite improvements in average concentrations, a number of areas still violate national air quality standards, particularly the ozone standard, and some of those areas do so quite often. In 1989, 370 counties, home to 95 million people, violated the national ozone standard at least twice (see Portney, 1990). Los Angeles, by far the worst offender, was in violation an average of 123 times each year from 1983 to 1985.

The link between ambient concentrations and health has proved difficult to identify and quantify. Some early research by Lave and Seskin (1977) could find no statistical link between ambient concentrations for pollutants from motor vehicles and impacts on health. More recent work by Krupnick (1988), however, finds that there are acute health improvements from decreases in ambient ozone levels in the northeastern United States. Krupnick also undertakes a benefit analysis and finds that these improvements are valued at an average of $500 per ton of VOCs reduced, with a range from $95 to $2,400 per ton.

The costs of standards in the 1970s have been estimated by White (1982) and by Bresnahan and Yao (1985), with the latter focusing on such nonpecuniary costs as decreases in driveability and performance. Bresnahan and Yao (1985) find that the nonpecuniary costs of the early regulations (1968 to 1973) were nearly the same order of magnitude as the pecuniary (capital, fuel, and maintenance) costs. From 1975 to 1981, on the other hand, the nonpecuniary component declined and automobile quality increased.

We can use Bresnahan and Yao's total cost estimates (pecuniary and nonpecuniary) and make a very rough comparison with Krupnick's benefit estimates. The estimated cost of meeting the 1975 emissions standards, which reduced the HC standard from 3.0 to 1.5 g/mi and the CO standard from 28 to 15 g/mi, is $1,305 per car (in 1988 dollars).[28] This is the present value of pecuniary and nonpecuniary costs over the life of the vehicle. A total of 8.6 million cars were sold in 1975. This means that the total cost of meeting the 1975 standard for all 1975 model year cars was $11.3 billion. Krupnick's $500 per ton estimate can be combined with the number of cars sold in 1975, the average number

---

[28]Bresnahan and Yao's estimate is $1,125 in 1981 dollars; I have used the Consumer Price Index (U.S. Department of Commerce) for "all commodities" to inflate to 1988 dollars.

of miles driven per car per year for an assumed 10-year life (see Stacy C. Davis and coauthors, 1989, table 3.8), and the change in emissions (1.5 g/mi) to obtain a present-value benefit estimate. This number is $647 million (in 1988 dollars). If we use Krupnick's highest benefit estimate of $2,400 per ton instead of the average $500, total benefits rise to $3.1 billion.

These numbers indicate that costs swamp even the highest benefit estimates. I hasten to point out, however, that the benefits measured are only those from acute health improvements from ozone reduction; no benefits from decreased chronic health problems or non-health effects are included. Also, no benefits from CO reductions are incorporated. In addition, Krupnick's benefit estimate is based on 1980s data, not 1975 data; the baseline year and corresponding baseline ambient concentrations can matter a great deal.

Nonetheless, the additional benefits would have to be enormous to bring total benefits in line with costs—between $8.1 billion and $10.7 billion, or 13 to 17 times the acute health benefits from ozone reductions. Moreover, my benefit estimate is overstated for two reasons. First, it assumes that the standard leads to HC emissions that are 1.5 g/mi lower in every year for 10 years; in reality, emissions systems deteriorate over time and emissions rise. Second, the estimate attributes benefits to attainment areas as well as nonattainment areas. Assigning the benefits only to nonattainment areas could decrease the estimate by more than 50 percent, which brings up a major point of this discussion: not only does it appear that the costs of the standards may have outweighed the benefits, but areas with no ozone problem have incurred a disproportionate share of these costs. Using the numbers above, attainment areas paid approximately $5.7 billion (half of $11.3 billion) for 1975 motor vehicle regulations that probably provided zero, or very few, benefits.

This last consideration is at the core of the federalism issues surrounding air pollution regulations. Suppose that my cursory benefit-cost calculation is off the mark and that the 1975 regulations did provide positive net benefits for nonattainment areas. Suppose further that these benefits would not have been reaped if the areas had established their own individual standards. In other words, the uniform standard provided a reduction in per vehicle costs due to economies of scale in production that were enough to bring about positive net benefits.

There is still no mechanism in place for compensating attainment areas, even though nonattainment areas would, in principle in this scenario, be willing to do so.

On the one hand, the CAAA of 1990 appears to improve the situation: only ozone nonattainment areas are required to meet the reformulated gasoline, RVP, and Stage II vapor recovery systems requirements; only CO nonattainment areas are required to meet the oxygenated fuel requirements; and only California is forced to begin using alternative-fuel vehicles in large numbers. These regulations do not apply to all areas of the United States. In addition, the Amendments allow states to adopt the tighter California standards, rather than abide by federal standards. In the past, states could not preempt the federal regulations.

On the other hand, emissions standards for all vehicles have been tightened and on-board diagnostic systems are to be required on all new vehicles. Given my admittedly crude benefit-cost assessment of early standards, I question the wisdom of these new requirements. Furthermore, the portions of the amendments that allow for differences among areas based on their attainment status are redundant. Many states are already doing these things. As of June 1990, 12 states had lowered RVP requirements to 9.5 or 9.0 psi; some apply the standard only in nonattainment areas and most only in summer months (*Clean Fuels Report*, 1990). Many nonattainment areas require service stations to have Stage II vapor recovery systems. Oil companies have voluntarily begun selling reformulated gasoline in many cities—although not with the exact specifications set out in the CAAA. And Colorado and other mountain states with CO problems have adopted higher oxygen requirements for gasoline sold during the winter months. It is difficult to see why any of these regulations must be established at the national level. Since states must develop State Implementation Plans to meet national ambient air quality standards, the incentive is already in place for them to institute regulations of this sort, or of whatever sort is needed to meet the standards. If they determine that it is not in their citizens' best interests, they do not institute the regulations. It is hard to argue that the federal government needs to intervene except to approve SIPs and enforce the air quality standards.

It might be argued that the clean-fuel vehicle requirements—both the fleet requirement and the California pilot program—are useful because they get an "infant industry" started. That is, vehicle and fuel costs might be lowered from the demonstration program and consumer acceptance of this new product might be increased. Whether or not these arguments have merit, many states were undertaking such programs on their own. The California Air Resources Board (CARB) adopted new emissions standards in October 1990 and required that alternative fuels must be made available at a certain number of service stations. The SCAQMD plan for the Los Angeles area requires the use of alternative-fuel vehicles. Many states have passed laws that alternative fuels must be used in fleets, both government and privately owned, or have provided subsidies and other incentives for private use of such fuels (see *New Fuels Report*, various issues).

Much of the cost of meeting the mobile source requirements of the CAAA—in particular, the costs of cars meeting the lower standards and on-board diagnostics requirements—will be borne by areas of the country that receive very little benefit. In our view, an experiment allowing attainment areas to establish a less stringent set of standards would be worthwhile. One possibility would be to allow attainment areas to maintain the current standards. This avoids the problem of numerous sets of standards while still reducing the cost borne by the cleaner areas.

Conclusions

That federalist systems of government have shortcomings as well as virtues is not a new observation. Although such systems have the potential for more efficient provision of publicly provided goods and services than unitary systems, this potential is often not reached. Moreover, the systems sometimes create their own problems.

In this paper, I have discussed the issues that need to be addressed when establishing the appropriate level of government for addressing energy security and environmental problems. I concluded that energy security issues are best handled at the national level, although some state policies may be warranted to address the economic dislocations of oil price shocks. I concluded that environmental problems, on the other hand, are best left to the states unless either spillovers of benefits and costs or economies of scale are prevalent.

My case studies of two federal government programs, the management of oil and gas resources of the Outer Continental Shelf and the regulation of motor vehicle emissions, indicate that these prescriptions may sometimes be too simplistic. In particular, the government action itself can create spillovers, and unless these spillovers are taken into account the resulting outcomes are likely to be inefficient. In its efforts to achieve energy security goals, for example, the federal government's OCS leasing policy imposes significant costs on coastal states. Since the states are uncompensated for these costs, they have the incentive to delay or halt OCS activities. By imposing national motor vehicle emissions standards, areas of the country that have little or no vehicle-related air pollution incur costs for very few benefits.

Problems such as these do not change my broad prescriptions. They do lead me to suggest, however, that federal policies take into account the differential regional impacts they sometimes confer. If some states or regions pay the costs of a program without reaping corresponding benefits, redistribution should take place to remedy that problem.

References

Anderson, Robert. 1990. "Reducing Emissions from Older Vehicles." Research Study #053, August (Washington, D.C.: American Petroleum Institute).

Bohi, Douglas R. 1981. *Analyzing Demand Behavior: A Study of Energy Elasticities.* (Baltimore, Md., Johns Hopkins University Press for Resources for the Future).

————, and W. D. Montgomery. 1982. *Oil Prices, Energy Security, and Import Policy* (Washington, D.C., 1985. Resources for the Future).

Bresnahan, Timothy F., and Dennis A. Yao. 1985. "The Nonpecuniary Costs of Automobile Emissions Standards," *Rand Journal of Economics* vol. 16, no. 4 (Winter) pp. 437–455.

Burtraw, Dallas, and Paul R. Portney. 1991. "Environmental Policy in the United States," in Dieter Helm, ed., *Economic Policy Towards the Environment* (Oxford: Blackwell Publishers).

Cicin-Sain, Biliana, and Robert W. Knecht. 1987. "Federalism Under Stress: The Case of Offshore Oil and California," in Harry N. Scheiber, ed., *Perspectives on Federalism* (Berkeley, Calif., University of California).

Clean Fuels Report vol. 2, no. 3, June 1990.

Cumberland, John H. 1979. "Interregional Pollution Spillovers and Consistency of Environmental Policy," in H. Siebert, ed., *Regional Environmental Policy: The Economic Issues* (New York: New York University Press).

Davis, Evan H., Scott T. Grusky, and Feriedoon P. Sioshansi. 1989. "Auto Emission Taxation: Alternative Policies for Improving Air Quality," paper presented at 11th Annual North American Conference of the International Association for Energy Economics, Los Angeles, Calif.

Davis, Stacy C., Deborah B. Shonka, Gloria J. Anderson-Batiste, and Patricia S. Hu. 1989 *Transportation Energy Data Book: Edition* 10, ORNL-6565 (Oak Ridge, Tenn., Oak Ridge National Laboratory).

Elliott, E. Donald, Bruce A. Ackerman, and John C. Millian. 1985. "Toward a Theory of Statutory Evolution: The Federalization of Environmental Law," *Journal of Law, Economics, and Organization* vol. 1, no. 2 (Fall) pp. 313–340.

Forecast '88/89: The Santa Barbara Economic Outlook. 1988. (Santa Barbara, Calif., Department of Economics, University of California).

Hamilton, Alexander, John Jay, and James Madison. |1787| 1961. *The Federalist Papers* (New York: New American Library).

Khazzoom, Daniel J. 1989. "The Use of Energy Conservation as a Strategy for Reducing Emissions in the Transportation Sector," testimony before California Energy Commission Proceedings on the 1989 Fuels Report (June).

Kriz, Margaret E. 1989. "Ahead of the Feds," *National Journal* (December 9).

Krupnick, Alan J. 1988. Chapter in *An Analysis of Selected Health Benefits from Reductions in Photochemical Oxidants in the Northeastern United States*, Brian J. Morton, ed. Final Report prepared for U.S. Environmental Protection Agency, Office of Air Quality Planning and Standards, Research Triangle Park, N.C. (Washington, D.C., Resources for the Future).

Lave, Lester B., and Eugene P. Seskin. 1977. *Air Pollution and Human Health* (Baltimore, Md., Johns Hopkins University Press for Resources for the Future).

Mason, Alpheus T. 1972. *The States' Rights Debate*, 2d ed. (New York: Oxford University Press).

Miller, Daniel S. 1984. "Offshore Federalism: Evolving Federal-State Relations in Offshore Oil and Gas Development," *Ecology Law Quarterly* vol. 11, no. 3.

Mitchell, Robert Cameron, and Richard T. Carson. 1989. *Using Surveys to Value Public Goods: The Contingent Valuation Method* (Washington, D.C., Resources for the Future).

Musgrave, Richard A., and Peggy B. Musgrave. 1976. *Public Finance in Theory and Practice*, 2d ed. (New York, McGraw-Hill).

New Fuels Report, various issues.

New State Ice Company v. Liebmann, 285 U.S. 262, 311 (1932).

Portney, Paul R. 1990. "Air Pollution Policy" in Paul R. Portney, ed., *Public Policies for Environmental Protection* (Washington, D.C., Resources for the Future).

———, David Harrison, Jr., Alan J. Krupnick, and Hadi Dowlatabadi. 1989. "To Live and Breathe in L.A.," *Issues in Science and Technology* (Summer) pp. 68–73.

Reagan, Michael D. 1972. *The New Federalism* (New York, Oxford University Press).

Rose-Ackerman, Susan. 1981. "Does Federalism Matter? Political Choice in a Federal Republic," *Journal of Political Economy* vol. 89, no. 1 (February) pp. 152–165.

Secretary of the Interior v. California, 52 U.S.L.W. 4063, 1984.

South Coast Air Quality Management District [SCAQMD]. 1988. *Draft Air Quality Management Plan* (Los Angeles, CA: SCAQMD) December.

Tiebout, Charles M. 1956. "A Pure Theory of Local Government Expenditures," *Journal of Political Economy* vol. 64 (October) pp. 416–424.

Tippee, Bob, and Bob Williams. 1990. "Operators Boost Offshore Action Where Leases, Permits Available," *Oil and Gas Journal* (June 4) pp. 64–72.

U.S. Congressional Budget Office. 1988. *Environmental Federalism: Allocating Responsibilities for Environmental Protection* (Washington, D.C., Congress of the United States).

U.S. Department of Commerce, Bureau of Economic Analysis. *Survey of Current Business*, various issues.

U.S. Department of the Interior, Minerals Management Service. 1986. *Managing Oil and Gas Operations on the Outer Continental Shelf* (Washington, D.C., U.S. Department of the Interior).

———. 1987. *5-Year Leasing Program Mid-1987 to Mid-1992, Proposed Final* (Washington, D.C., U.S. Department of the Interior).

Walls, Margaret A. 1991. "Dynamic Firm Behavior and Regional Deadweight Losses from a U.S. Oil Import Fee," *Southern Economic Journal* vol. 57, no. 3, (January) pp. 772–788.

White, Lawrence J. 1982. *The Regulation of Air Pollutant Emissions from Motor Vehicles* (Washington, D.C., American Enterprise Institute).

Williams, Bob. 1990. "Point Arguello Project Start-up Blocked Again," *Oil and Gas Journal* (November 19) pp. 34–35.

Can Electric Power—A "Natural Monopoly"—Be Deregulated?

VERNON L. SMITH

The overwhelming event in the political economy of the 1980s has been the worldwide revolt against the central control and regulation of economic activity by governments. Neither national boundaries nor particular political systems have escaped these forces as both democratic and totalitarian regimes have succumbed intellectually to classical liberal concepts of economic organization based on decentralization. This is nowhere better illustrated than in the following background statement of the report of the task force on electric power privatization appointed by one of the first of the twentieth century's democratic socialist governments—New Zealand.

> This review of the electricity industry is part of a wide-ranging re-orientation of industry policies within New Zealand. While the details of deregulation vary considerably from industry to industry, the key broad objective is to maximize national income by encouraging greater efficiency in resource use. This objective is being pursued through a much greater reliance on market incentives, and a re-defining of the role of Government to that of establishing a legal framework within which individual and commercial decision making can take place, determining welfare transfers, and, where necessary, arranging the provision of public goods.

This paper was presented in the John M. Olin Distinguished Lectureship Series at the Colorado School of Mines on October 22, 1991.

This process has led to a number of specific initiatives being taken for a variety of industries, including for example:

- removing legislative barriers to entry to increase competitive pressures;
- clarifying price signals by easing various protective mechanisms, reducing subsidies, and removing unnecessary regulations;
- ensuring a competitive environment that is as neutral as possible, both within and between industries, and avoiding the Government discrimination between different forms of activity;
- providing a more stable and predictable economic policy environment; and
- placing more emphasis on generalized trade practice legislation and reducing the emphasis on industry-specific regulation. (Electricity Task Force, 1989, pp. 9–10).

While this statement typifies the new openness of New Zealand to classical economic prescriptions, the idea that such concepts can be implemented in the context of the electric power industry is not typical of classical and contemporary economic views of electric power as inherently a natural monopoly industry. How is it possible that this industry might be reorganized along lines "ensuring a competitive environment"?

The theory of natural monopoly seems to have originated with John Stuart Mill, who argued that it would be uneconomical and duplicatively wasteful for cities to be connected by parallel railroad tracks, or for a city to be served by more than a single postal service (Mill, 1848). Although Mill never coined the term, his arguments evolved into the theory of natural monopoly, which became standard in economics textbooks for decades, down to the present. According to this theory, an industry is a natural monopoly if the entire market demand can be served at lower cost by a single firm than by two or more firms (Sharkey, 1982). A sufficient condition for natural monopoly is the existence of increasing returns to scale (that is, declining average unit cost) for output levels at least as large as the quantity demanded at a price that covers average cost. Although this is a static theory it is thought to apply with particular force to the quintessentially dynamic environment of the electric power industry. In electric power, scale economies in high voltage transmission and in generation have traditionally been invoked to justify regulation or state ownership. But the existence of multiple parallel transmission and generation units owned by the same firm makes it plain that the traditional static argument for natural monopoly is contradicted by observation.[1] A more compelling argument rests on the existence of networking and coordination economies in the generation-transmission nexus; these economies are obtained through the use of a computerized dispatch center to load multiple generation units dispersed in space on the transmission network of the typical large integrated electric utility.

In this paper I propose to articulate a view of how a privatized, deregulated, competitive electric power regime would function and to sketch the broad outlines of the structure needed to achieve these objectives. Although the market for these ideas is politically remote in the United States, this is not the case internationally. The United Kingdom has privatized its electric power industry under a regime explicitly designed to avoid the high cost of American-style rate-of-return regulation, and, as noted above, New Zealand has similar objectives for their state-owned electric power industry. Similar movements are under way in many other nations worldwide. Although this paper is written from the perspective of privatizing a state-owned power industry, it also represents an articulation of the end objectives proposed for a deregulated industry.

The key factor that has made these developments possible is the computerized dispatch and coordination center, which allows the coordination advantages of central control to be combined with the information advantages of a decentralized auction bidding market. But possibility is not reality. The necessity for political compromise with the forces protecting the income distributional consequences of incumbent industry structure can short-circuit the productivity potential of more disciplined market-based forms of organization. This is evident (as I write) in the early months of the operation of the newly privatized electric power system in the United Kingdom, which is already showing price inflexibilities because of political obstacles to the creation of a

---

[1]Scale economies for both generators and transmission lines are exhausted at capacities near the frontier of the design and construction knowledge for building large scale units. Beyond a certain capacity such knowledge is uncertain and experimental, which leads to rising capital costs per unit of installed capacity. Consequently, total average costs increase beyond these capacity levels, which are small relative to demand. At this point one installs multiple parallel units. Thus fossil fuel power plants typically consist of two or more turbine-generator units served by a common fuel supply system and common interconnection buses.

greater number of competing power generation units. Similar forces may be at work in New Zealand and other countries.

Principles and Guidelines

Defining Rights

The key issue in institutional design for the deregulation of electric power is in the specification of property rights. By "property" rights is meant rights to act, that is, human decision-making rights, not rights only to physical property. In order that institutions can function without continuous intervention by government, rights to act must be well-enough (if not always perfectly) defined, allowed to become valuable, and be transferable (alienable by trade, gift, or bequest) so that their value can be conditioned and tested by the costs of alternative resource uses. Rights holders must bear the costs—and receive the benefits—consequent to their actions. To achieve this in natural monopoly situations requires several state-specified modifications of the customary accouterments of property.

With electric power networks it is of crucial importance in conceptualizing rights to avoid identifying such rights with the physical character of the network. That is, competition must derive from a contest between multiple rights holders, not from the multiplicity and divisibility of physical facilities, which economy may require us to avoid. Rights, and the actions they permit, are required only to be feasible physically; the rights themselves do not have to be defined in terms of physical flows or representations. Thus, if B sells power to A, there is no requirement that each can identify a distinct flow of energy from B's location to A's location. Thus, A might contract with B, a downstream source, whose output frees upstream power for A's consumption. It is important, therefore, for conceptual clarity, and for understanding what we are about, to avoid interpreting rights in terms of the physical complications of the network, its users, and how they are physically served. That will simply make it difficult to understand and appreciate the property right issues that have to be addressed.

Finally, rights must be expandable (and contractible), either through free individual investment or joint investment; that is, entry and exit must be unconstrained by artificial legal barriers. Consequently, the design must be responsive to dynamic change and evolution and provide incentives for this response to reflect costs and values.

Divisible Rights, Indivisible Facilities: An Example

These abstract principles can be conceptualized more clearly if we examine a simple concrete example. Suppose that a product requires a physical channel or facility, such as a pipeline or transmission line, and demand is such that a single operating facility is adequate to satisfy peak requirements. Such are the classic conditions for natural monopoly. The traditional response is that the facility will have one owner—either the state or a private company whose prices are regulated by the state to yield no more than a "fair" rate of return on investment.

Instead, suppose the state simply specifies the following set of property right rules that must govern a co-tenancy contract among competing independent co-owners of the facility (Braman, 1990; Smith, 1988).

1. Each co-owner acquires capacity rights to the facility in proportion to his contribution to capital costs and any output-insensitive operating costs.
2. These rights are freely transferable. They can be sold, leased, or rented to any outsider or to any co-owner, subject only to the usual antitrust provisions that apply to any other industry.
3. The co-owners have some internal agreement for sharing variable costs whenever rights are exercised. For example, variable cost might be shared in proportion to the capacity utilized, up to each co-owner's share of capacity. (See Moulin and Shenker [1992] for optimal sharing arrangements when unit costs are not constant.)
4. The facility is managed as a cost center by a separate operating company; it has no marketing function—this is performed by the competing co-tenants.
5. Any co-owner or any outsider can increase his share of capacity utilization rights by expanding capacity unilaterally.

Co-tenancy contracts, or joint production ventures, are not new; they are well established even in the electric power industry. What is relatively new in the above rules is their reconstitution as an instrument of competition for providing buyers of the product or service of the single facility with a choice among alternative sellers. Co-tenancy contracts in practice normally contain rules similar to rules 1 and 3, but rules 2, 4, and 5 are considerably different. For example, instead of rule 2 existing such contracts prevent the sale of a co-owner's capacity rights without first giving the other co-tenants the right of first refusal. In place of rule 4, for example in co-owned generators, it is customary to appoint one of the co-owners (usually the one with the largest share of capacity) to act as the manager of the operating company with provision for sharing the costs of that management. Finally, instead of rule 5, it is usually provided that capacity expansions require approval from, and some agreement about sharing cost among, all the co-owners. The rules as stated above are designed to remove provisions, such as the ones just enumerated, that are likely to be anticompetitive and to provide for capacity expansion in response to market signals.

In this example it is worth emphasizing that short-run competition among the co-tenants to market the product using the services of the fixed facility depends upon each (or at least one) co-tenant's having *excess capacity* rights. Thus, it is because cotenant A is serving fewer customers than she has a capacity to serve that cotenant A can solicit customers away from cotenant B. (This point, while perhaps obvious in the case of a pipeline, or a co-owned generator, also must be understood in applying the co-tenancy concept to distribution, which I undertake below.) Therefore, the ability to compete for existing customers, and the ability to serve net aggregate growth customers, depends upon having a competitive expansion rule. This is why in rule 5 we allow any incumbent co-tenant, or any outsider, the right to unilaterally expand capacity. (For a more formal treatment of these issues see Alger and Braman [1991]).

Shared Capacity Systems in the Economy

There are many examples in business practice in which the contractual sharing of capacity rights has arisen spontaneously for various reasons,[2] not usually for the explicit purpose of creating competition, although that is sometimes an unintended by-product.

---

[2]There are also numerous examples in history. Every historical arrangement for sharing a common property resource is an indirect example, since the co-tenancy or joint venture contract creates a common property resource by agreement. For historical examples, and an analysis of why some arrangements have worked and others have failed, see Ostrom (1990).

It is common in the United States, and elsewhere, for morning and evening newspapers in the same city to share the ownership and use of the specialized printing facility each requires for publication. It is clear in this example that printing machines are a capital input the sharing of which has no necessary connection with the vigor with which two newspapers compete for subscribers, sales, advertising, and editorial influence, and may enhance such competition by economizing on production resources. This example also illustrates the need for the state to require such co-tenancy contracts to be competitively ruled; such contracts have many times been embellished, by competing newspapers, to include marketing issues such as the pricing of advertising space. Although the courts had earlier disallowed such provisions, since 1970 newspaper operating agreements have been exempt from antitrust action.[3]

Shopping malls are a particularly innovative example in which economies of scope, scale, and agglomeration are captured in a single facility in which private space rights are combined with the sharing of a wide variety of common inputs: parking, walkways, lounge facilities, utilities, security, and custodial services. The resulting economies and input sharing combine to enhance greatly the contestability of retail distribution.

Finally, investor-owned utilities in the United States often own pipelines, transmission lines, or baseload generation units under co-tenancy contracts. Although, as noted above, such contracts often include undesirable provisions, which should be excluded if they are to be competitively ruled, it is clear that such arrangements are workable and even attractive to the participants. They are used to share risk and spread capital cost, but often they result in competition. Thus, two power companies sharing capacity rights to a baseload generator may both sell power in spot energy exchanges, or under long-term contracts, to third parties. Since the power from the co-tenancy unit is injected into their

---

[3]Such exemption is directly contrary to the principle that such agreements must be competitively ruled. See Reynolds (1990) for a review and evaluation of the public policy decisions in one court case and in the Newspaper Preservation Act of 1970, which legislated antitrust exemption for newspaper joint operating agreements. The court did not rule against newspaper joint operating agreements, but did rule against the price-fixing and profit-pooling provisions of such agreements. The court's remedy in this case corresponds to the principles we propose here for governing co-tenancy agreements.

respective networks, it may become part of these types of competition between the two firms. Similarly, sharing capacity rights to a transmission line has sometimes afforded a third party, needing to acquire capacity rights to the line, two sellers with whom to negotiate.

Avoiding Regulatory Problems

All property rights systems involve "regulation" in the sense that what decision makers can or cannot do is constrained by the rules governing rights to act. The regulatory problem we seek to avoid is one in which continued intervention or monitoring by the state is made necessary because the announced intentions of regulation—such as setting prices to limit profit to a "fair" return on the "prudent" use of capital—create incentives that are incompatible with the intentions. Thus, firms may see rate-of-return regulation as providing a guaranteed profit markup on cost and may be poorly motivated to control costs, unlike in alternative regimes in which a residual claimant gets to keep whatever is saved. Prices, under such regulation, are set by adding capital cost and a profit rate to other costs and thereby reversing the competitive process in which prices determine the amount of capital cost one can afford to undertake to make a profit (if any).

Anyone seeking to avoid these problems with a property rights approach is necessarily charting a course that cannot promise certainty of outcome. But the potential improvement is surely worth the risk, given the proven gross inadequacies of state ownership and of regulatory regimes.

Defining Private Rights and Private Entities

There are three users whose contribution and rights to an electric power network require delineation: generation, industrial companies, and retail power merchants, which we will refer to as gencos, indcos, and retailers, respectively. These users would be co-tenant owners or rights holders in a transmission network company, which we will call Transco; Transco is an operating subsidiary co-owned by the users. We will also consider a structure for distribution, which is parallel to that for transmission, in which power is marketed to final consumers by retail merchants who co-own their local distribution network as an operating subsidiary.

Generation

Base and Reserve Rights.   All gencos would have location-specific power *injection* rights to the network. Gencos are for-profit private corporations who compete to sell energy through the grid to discos and indcos. Gencos purchase initial endowments of *base* rights to the grid. These are Transco network rights to sell power anywhere to wholesale buyers, provided only that reserve rights are not required to serve those buyers. Base rights extend automatically to genco co-tenant owners of the transmission network. Such rights are paid for by membership fees proportional to each genco's capacity to inject power into the grid: lump-sum payments initially and an annual fee thereafter for maintenance of grid capacity. The latter is paid to Transco to cover maintenance, operating, and overhead costs. Some adjustment of these fees to take account of "remoteness," using average transmission penalties, may be desirable. But in state-owned power industries the sunk cost character of the grid argues against historical accuracy in this allocation. Whenever new generation capacity is added, either by existing gencos or by new genco entrants, additional base rights to the grid must be acquired by purchase in the same way that such rights would be obtained by gencos formed from the initial divestiture.

At their option, individual gencos also may be awarded *reserve* rights (Read, 1991), which are capacity rights on particular transmission lines subject to congestion and operation under capacity constraint.[4] Rights on such lines would be reserved for those buying them, who contract to ship power beyond nearby distribution buses. They entitle the holder to receive capacity rents at the congestion prices for all periods in which the indicated lines are loaded to capacity. Holders also receive the node-to-node price differences based on incremental transmission loss on reserve lines. Reserve rights would be perpetual but alienable. The putative purpose of reserve rights is to establish priority and entitlement

---

[4]In New Zealand the two current examples of potentially constrained lines are the high-voltage DC line and the Auckland Isthmus AC line.

to congestion prices on lines subject to constraint so that clear and appropriate signals can be provided for new capacity investment. New reserve rights on existing reserve capacity lines are obtainable at any time for the account of any genco or any outside company by construction of new capacity. The concern that reserve capacity obtained by expansion might introduce monopolistic elements has led to the suggestion that such rights should have a time limit that expires like a patent. Because of the hazards of choosing the appropriate time period one might simply require the rights to be sold at auction after five years to at least two parties. This would subdivide ownership but improve the return and therefore the incentive of the original investor.

If nodal and congestion prices are found to provide inadequate incentives for new transmission investment, reserve right holders could be given the right to levy wheeling charges. The dispatch center would treat these as if they were additional incremental "losses" and reallocate loads appropriately. Ideally, those gencos and retailers that stand to benefit from capacity expansion would form coalitions to make efficient capacity expansions, but the right to make unilateral expansions, and charge appropriately, is needed to discipline any such joint action.

Contract and Spot Selling Rights. Gencos would be free to forward contract with retailers and indcos to serve their power needs. Literally, power would be injected at the genco unit's location, and simultaneously power would be withdrawn at the wholesale delivery bus. The actual pattern of power flows would occur in response to the central economic dispatch program of the Transco coordination center. Such contracts define the financial arrangements and are consistent with physical delivery and production in a node-to-node representation of transmission losses.

Alternatively, gencos could sell power in half-hourly, or more frequent, spot auctions as institutional and technological developments permit. Although it is likely that gencos and their wholesale buyers will rely initially on contracting for power transfers, based on deregulation experience in other industries (such as natural gas) increasing reliance on the spot market is likely as agents become familiar and comfortable with their operation. Gencos accomplish this by submitting to the dispatch center a willingness-to-accept offer supply schedule for power at their injection buses. The center then programs a particular production level off this schedule, depending on the optimal dispatch pattern. Even with forward contracting, power not taken by the buyer can be dispatched spot under the control of the buyer or the genco as the terms of their contract specify. Thus, even with forward contracting, spot exchanges are certain to be common as a means of adjusting demand realizations across time and space to supply.

Retail Power Merchants; Distribution Companies

Base Rights. As co-tenant contract owners of Transco, retailers have *base* capacity rights to the grid, which allow them to withdraw power from the network at their delivery buses. These rights are paid for by membership fees proportional to the retailers' capacity to withdraw power from the grid: lump-sum payments initially, annual fees thereafter for network maintenance, operating, and overhead costs (perhaps location-adjusted for remoteness). Capacity expansion for new customers would require the purchase of additional capacity rights to Transco. Retailers could also acquire reserve rights at auction on potentially congested constrained segments of the grid. This would allow them to contract with gencos that do not have reserve rights and collect revenues from congestion prices when the reserved lines are up to constraint and revenues based on incremental loss prices when the lines are not constrained.

Contract and Spot Purchasing Rights. Retailers may forward contract with gencos, but would be free to satisfy any part of their power demand by purchases in the spot auction. They accomplish this by submitting willingness-to-pay bid demand schedules to the dispatch center for power delivered to their respective load buses. Based on economic dispatch the center accepts a load commitment on this bid schedule. This should be decentralized because the retailer knows best the nature of its interruptible or load-shedding contracts, and the price levels above which it will want to invoke the load reduction terms provided in these contracts. Obversely, exposure to half-hourly peak load prices at their load buses will provide retailers with both the incentive and the cost input data to further fine tune and write demand management contracts with individual rate payers.

141

The character of demand management is certain to evolve and become increasingly sophisticated and price responsive once the precedent is established that local distribution is exposed to time-variable spot prices. On peak these prices are capable of reaching very high levels as high-cost, but flexible, gas turbines or other fossil fuel power sources are switched on line to produce to demand. One can expect such spot pricing of wholesale power to induce changes in residential charges much as the hotel/motel industry takes for granted time-of-week and seasonally variable room rates. This in turn will transform investment behavior since capacity is determined by peak rather than off-peak demand in industries (such as hotel/motel, airline, and electric power) that must satisfy instantaneous demand and cannot inventory the services rendered. Retail responses to time-variable prices are limited by the fact that most customers are not metered half-hourly. But spot pricing will give retail distributors an incentive to move expeditiously to seasonally (monthly, weekly) variable rates and to encourage the installation of more time-of-day metering and rate schedules.

Even if power is purchased primarily by forward contracting, power contracted for but not utilized at particular times is available for dispatch in the spot market. Depending on the nature of these contracts, retailers (or gencos) may offer to sell the surplus power by submitting the appropriate offer supply schedule to the dispatch center.

Power Merchants as Co-owners of Distribution Networks.
Local distribution networks are technically small-scale, low-voltage versions of the transmission network with dense radial or looped (sometimes networked) connections for power takeoff to residential and commercial customers.

The co-tenancy principle would be applied at the local level in seeking to solve the problem of introducing competition in the provision of local distribution services. Capacity rights to each distribution company would be held by two or more retail power merchants. Only the power merchants, and *not* the distribution company, would be in the business of marketing electrical energy under this regime.

To see how this might function, think of a large local distribution company as consisting of two departments: (1) customer service and billing and (2) operations and maintenance. Now, spin off customer service and billing into two or more retail power companies and spin off operations and maintenance into a separate management company co-owned by the retail merchants under a competitively ruled co-tenancy contract. Each merchant is assigned capacity rights (retail customer hookups) to customers, most active, some inactive (or not yet connected). Although each merchant may initially have customer accounts in the same or contiguous geographical areas, competition would be achieved by allowing inactive capacity rights to be exercised anywhere in the distribution system.[5] Thus, no merchant would have a geographical local service monopoly. Any merchant could acquire additional capacity rights by purchasing the inactive capacity rights of another merchant or by investing in new capacity, that is, installing new distribution transformers, lines, and connections to new geographical areas under development by builders. But such expanded rights could be exercised anywhere in the distribution area, so the customer accounts in a particular new housing development might be won in competition by any retailer with inactive account capacity. New power merchants can enter the distribution co-tenancy either by purchasing capacity from incumbent merchants or by building new capacity.

Since power merchants share the cost of the local network, each merchant would recover these costs from retail accounts as he saw fit in competition with other merchants. The merchant's location-specific bids for power would be aggregated at the delivery bus connecting Transco with the local distribution company. Since these buses are typically metered half-hourly (Electricity Task Force, 1990, p. 95), the aggregate purchases of all merchants at the delivered spot price are known and the spot price is the same for everyone taking power at a given load bus. Retail power merchants could compete for customer accounts on the basis of different price structures: various combinations of fixed versus variable rates, flat rates, time-of-day charges for the larger customers, and time-of-day combined with load management

---

[5]Not all customer hookups are created equal. Lines and poles in foothill terrain are more expensive to install and maintain than those on flat rectangular subdivisions. Similarly, high-density connections to apartment buildings are less costly than those to private residences. It follows that practical implementation of the concept of capacity rights to serve individual customers would need to identify and distinguish among a few major cost classes. Retail rates would then reflect these differing costs of capital and maintenance. Of course this is desirable in a user-pays economic environment disciplined by free choice.

systems controlled by the dispatch center as such technologies become available at lower cost in the future. Different power merchants might specialize in different types of metering and price structure services. Even simple watt-hour (or cumulative) metering at the residential level permits seasonally variable rates—monthly or bi-monthly, depending upon the frequency of reading.

Industrial Bulk Buyers

Indcos can purchase power from retail power merchants, in which case they use the local distribution network, and indco rights to the grid, like residential customers, are derived indirectly from power merchant purchases of base rights.

Alternatively, they have the right of bypass by taking power directly from the distribution companies' load buses served by Transco or through interconnections purchased or leased from distribution companies. In the bypass case, indcos would acquire location-specific base capacity rights to withdraw power from the grid, paying for such rights through membership fees proportional to capacity the same as a retailer. Power could then be purchased from gencos by forward contract, or by entering the spot auction with bid demand schedules submitted to Transco's dispatch center.

Transmission: Structure, Operations, Pricing

Ownership

Transco would become an operating company run as a not-for-profit cost center co-owned by all entities with base injection or withdrawal rights to the grid. Ownership shares and the financial obligations of users are in proportion to their capacity rights. Although it has cost-minimizing commercial objectives it is more like a club, cooperative, or rule-governed contractual arrangement than it is a private corporation entirely on its own in making its rules. It is a property right creature of the state differing from the corporate enterprise seeking a return on its investment. It is not an investment company that makes capital expansion decisions. Those decisions are made by its parent users or outsiders who buy into the co-tenancy contract by purchasing capacity rights and making decisions to expand capacity in distribution, generation, or transmission.

Transco will collect revenues. These revenues are in the form of annual fees, proportional to user capacities, to defray the cost of main-taining the grid, operating the dispatch center, improving and expanding the metering of power, investing in communication and computer facili-ties for coordination, and providing load flow information to users. The second major source of revenue is derived from node-to-node pricing of power flows. Because these prices are at incremental transmission losses (marginal transportation cost) that are approximately twice the average unit losses, there is an operating profit approximately equal in amount to the actual average cost of lost power on each leg of the network.[6] (This ignores the small fixed losses to be prorated.) This revenue, which is col-lected from all end users in the form of location-adjusted transportation prices, is appropriately paid out to the holders of base rights in accordance with their ownership shares. The revenues on reserve right lines would be paid as a dividend only to reserve right holders.

A minor source of revenue would be special charges for load flow simulations conducted by the center as a paid-for service for users contemplating new investment in distribution, generation, or transmis-sion capacity and needing such information for locating and designing new capacity.

Control Center Operations

The core of transmission operations is the dispatch or control center. In a fully developed spot market, the center is responsible for economic dispatch, that is, selecting generator load levels to meet wholesale demand so as to maximize the gains from exchange (gross surplus) as revealed in the offer supply schedules of generator owners, the bid

---

[6]Power losses on a transmission line of given size and electrical resistance vary approximately in proportion to the square of the power delivered. Consequently, the marginal cost of lost power is approximately linear and twice the average cost of lost power per unit delivered. That is, if power loss is $P_L = \alpha P^2$, where α is the loss coefficient and P the power delivered, then marginal cost is $2\alpha P$ and average cost is αP. Under marginal cost pricing, unit profit is then $2\alpha P - \alpha P = \alpha P$.

demand schedules of wholesale buyers, and the electrical characteristics (impedance and transfer capacity) of each segment of the transmission network. Node prices are computed as a by-product of the maximization algorithms that allocate load demand among the generator units and power among the wholesale delivery buses.

It is worth emphasizing that generator owners are on their own in selecting the terms on which they are willing to supply different amounts of power. There is no artificial requirement that the generator supply schedule conform to a generator's incremental heat rate (marginal cost of fuel) as defined by the engineer. Gencos are free to submit offers that are well above marginal fuel costs (if they think dispatch will accept such offers for local commitment) as a means of recovering capital and other fixed costs. They are free to offer power at less than marginal fuel cost if by so doing they expect to avoid a greater cost of shutdown and start-up. The dispatch center runs a spot market that maximizes the total buyer/seller subjective gains from exchange, not a generator loading rule that minimizes energy cost only. If generator revenue covers capital cost it is either because gencos earn it in the spot market or through contract sales. If power merchants resell surplus power spot and receive revenues high enough to cover their contract costs, it is because spot prices are high enough to cover unit-fixed as well as variable costs.

The center contracts for reactive power, and for voltage/frequency control, and perhaps for spinning reserve. A first priority of the dispatch center is to maintain the stability and integrity of the grid as a power delivery system. This is actually part of "economic dispatch" in the sense that disruption costs, greater than the savings from adhering to the strict short-run economy loading of generators, are incurred if voltage control and spinning reserves are inadequate to avoid brownouts and blackouts.

Spinning reserve can be supplied off the dispatch program of the center as an integral part of spot pricing. On the supply side, gencos submit lump-sum offer prices for minimum spinning generator capacity, plus the prices and quantities at which higher outputs would be supplied up to maximum capacity. Any generator accepted for load commitment at less than maximum capacity is providing part of the needed spinning reserve. The supply of spinning is articulated by the center accepting enough surplus on-line generator capacity (above instantaneous demand) to satisfy the required spinning reserve. The cost of spinning is thereby included in the pattern of nodal spot prices. Initially, spinning reserve is likely to be provided by genco contract with the center but can be shifted into the spot auction as the latter becomes developed.

Pricing

The dispatch center computes all node prices. At a power delivery node this is a spot selling price and at a genco production node it is a spot buying price. Thus, at node i, the price (of real power) is $P_i = p(1 - ITL_i)$ where p is the price at the reference node (usually the supply cost of the most expensive generator on line) and ITL_i is the incremental transmission loss charged to node i (equal to the system increase in transmission loss if one were to draw one additional unit of power from node i).

The marginal transmission cost of a unit of power flowing from node j to node i is $t_{ji} = P_i - P_j = p(ITL_j - ITL_i)$. Thus, wholesale buyers at delivery nodes pay the marginal supply cost of produced power plus the marginal transmission system cost of transporting power to their nodes. At this allocation equilibrium it is irrelevant to a buyer at node i where the power comes from. Literally, it comes from, or is influenced by, all sources in the economically efficient equilibrium calculated by the dispatch center. When power is purchased by contract, these same prices are used to calculate (or estimate) the transportation penalty on power from the contract supply node to the delivery node. The contract is a financial instrument specifying the transmission-adjusted cost paid for power injected at node j and withdrawn at node i.

Exception: Capacity Constrained Lines

An exception occurs in this pricing regime any time a line is loaded to its capacity. In the New Zealand network this tends to occur on the High Voltage Direct Current (HVDC) Line connecting the two islands. Now the system devolves into two independent pools. The congestion price at the North Island node (receiving power) now rises inelastically above the South Island price plus the marginal cost of transmission to the North node. Marginal units of power at North Island nodes necessarily come not from "everywhere" but from North Island generators; similarly marginal units at South Island nodes come from local generators as the system becomes bifurcated. Hence the logic of thinking in terms of the concept of reserve rights on the HVDC line if it is subject to congestion pricing. Reserve right holders receive the congestion-induced supernormal revenues on the HVDC line.

Accountability of the Transco Operating Company

Co-tenancy property rights regimes are more cost-efficient the smaller the number of right holders. This is because each experiences a signifi-cant portion of the cost of the operating company and is motivated to exert discipline on the company's cost control. This argues strongly against atomized gencos and retail merchants. There is a trade-off between the competitive gain from more co-tenants and associated (unknown and unmeasured) losses in cost discipline. But there is one external source of market discipline of these costs: power in the long run must compete with gas and other energy sources, and the cost of organizational slack in transco will diminish this competitive capacity.

Summary

Figure I provides an overall perspective on the proposed structure of a deregulated electric power industry. Customers are served by compet-ing retail power merchants among whom individual customers are free to switch subject to any start-up charges levied by individual retailers of power. Retailers acquire capacity rights (hookups) to serve individual customers by investment in such capacity in the distribution company to which their customers are connected. Each distribution company is an operating subsidiary jointly owned by two or more local retail power merchants; it is operated as a cost center and has no marketing rights. Retailers pay the operating costs of distribution, which are prorated in proportion to customers served by the retailer. Retailers also pay the costs of power loss in the low-voltage local distribution network; such costs are prorated in proportion to total power con-sumed by each retailer's customer accounts. Retailers submit to the dispatch center bid schedules for power delivered to the interface bus

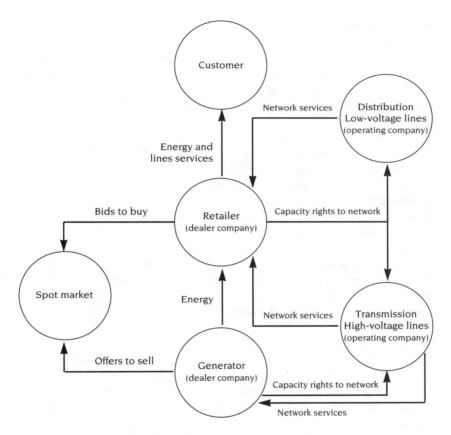

Figure 1. Contractual relationships and the spot market.

connecting the local distribution company with the transmission net-work. In each time period (for example, half hourly) the dispatch center chooses a single price at which the aggregate of these bid demand schedules is met by system supply at each delivery bus. This price includes the incremental loss of power in transmission that is charge-able to each delivery point. Retailers acquire power drawing rights from the grid through their initial investment in the grid, and pay their share of the operating costs of the network and the coordination and dispatch center. Retailers compete in pricing energy and lines service to cover all these components of cost. In this process they are free to seek specialized niches in which some customers might be charged flat rates per kilowatt consumed, others a fixed charge plus a flat rate, and still others time-of-day prices coupled with load management systems. Retail merchants acquire new hookup capacity rights either by expanding capacity or purchasing unused capacity rights from other merchants. No retailer or local distribution system has a local area franchise monopoly.

Transmission consists of the high-voltage lines business and, together with its coordination center, is co-owned by the retail distributors and the generating companies. It is operated as a cost center, with no marketing rights. Symmetrically with retailers, generator companies purchase base rights to the grid in proportion to their capacity to inject power into the network. Generation companies submit location-specific offer schedules expressing the terms on which each is willing to supply power to the grid. The dispatch center uses an optimization algorithm for allocating system load among all generators and simultaneously computes node-to-node prices so that system monetary surplus is maximized. Power intake nodes are priced at marginal supply cost, and delivery nodes are priced at marginal willingness to pay. The difference between intake and delivery node prices reflects the marginal cost of lost power in transmission when no transmission line is loaded to capacity. In addition to this wholesale spot auction in power, retailers and generators are free to engage in forward contracting, with any surplus or deficiency in power requirements satisfied in the spot auction. Reserve rights on lines likely to be loaded to capacity can be auctioned to the highest bidders in a uniform price auction. Reserve rights holders would then be entitled to all revenues earned on such lines based on incremental loss prices and congestion prices when a line is loaded to capacity. The purpose is to provide incentives for investment in increased capacity on such lines.

The dispatch center is also responsible for the stability of the transmission network and is empowered to contract with generator owners to commit their generators to such ancillary services as voltage and frequency control, producing reactive power and providing spinning reserve unless the latter is provided directly off the dispatch center's generator loading algorithm.

The putative purpose of these arrangements is to provide mechanisms whereby all users pay the opportunity costs occasioned by their consumption and production decisions and the resulting prices provide signals for new investment that allow capital costs to be weighed against the value of increased capacity.

Acknowledgments

I am indebted to Penelope Brook and Stephen Jennings of C. S. First Boston, New Zealand, and to Susan Braman, Grant Read, and many others working with the New Zealand government on the innovative efforts of that country to privatize their electric power industry while avoiding the pitfalls of heavy-handed regulation. Through their efforts I learned much about the possibilities and mechanisms whereby electric power in New Zealand might be reconstituted as a competitive industry.

References

Alger, Dan, and Susan Braman. 1991. "Competitive Joint Ventures and the Regulation of Natural Monopolies," draft (Washington, D.C., Economists Incorporated).

Braman, Susan. 1992. "Theory and Applications of Competitive Joint Ventures" (Ph.D. thesis, Georgetown University, Washington, D.C.).

Electricity Task Force. 1989. *Structure, Regulation and Ownership of the Electricity Industry*. Report of the Electricity Task Force (Wellington, New Zealand, September).

Mill, John Stuart. 1848. *Principles of Political Economy* (reprinted 1900, London, Colonial Press).

Moulin, Herve, and Scott Shenker. 1992. "Serial Cost Sharing," *Econometrica* vol. 60, no. 6, pp. 1009–1038.

Ostrom, Elinor. 1990. *Governing The Commons* (New York, Cambridge University Press).

Read, E. Grant. 1991. *Transmission Pricing Issues for New Zealand*, Status Report to Transpower Establishment Board (February 28).

Reynolds, Stanley S. 1990. "Cost Sharing and Competition Among Daily Newspapers" (Department of Economics, University of Arizona, October).

Sharkey, William W. 1982. *The Theory of Natural Monopoly* (Cambridge, England, Cambridge University Press).

Smith, Vernon L. 1988. "Electric Power Deregulation: Background and Prospects," *Contemporary Policy Issues* vol. 6, no. 3, pp. 14–24.